Finite Groups

A Simple Introduction

by

Dennis Morris

Copyright: Dennis Morris

dennis355@btinternet.com

All Rights Reserved

Published by: Abane & Right

56 Coach Road

Brotton

Saltburn-by-the-Sea

TS12 2RP

01287 678918

September 2016

Contents

Preliminaries 1
   The numbers of the finite groups to order 100 1
      Index to tables: 1
      The simple finite groups: 2
      The tables: 2
      Proper subgroups to order 15: 3

Introduction 5
      Finite groups are more than only symmetry: 5
      It's the concepts that count: 6
      Permutation matrices: 6

The Permutation View of Finite Groups 8
      Cayley's Theorem: 8
      The permutation view of finite groups: 9
      A notation for permutations: 9
      Multiplication of permutations: 10
      An abstract view: 11
      Mappings: 11
      Back to the permutation view of finite groups: 12
      Finite Groups: 12
      The group order and why finite: 13
      Squares and triangles: 14
      Summary so far: 15

Products of Permutations 17
      What is a permutation?: 17
      Closure again: 18
      Identity: 20

Powers of permutations: 20
The order of a permutation: 21
Non-commutativity of permutations: 21
The multiplication of permutations: 22
Another example: 23
Warning: 23
Inverse permutations: 24
Cycle structure and a note on notation: 25
The order of a permutation from the cycle structure: 26
Summary: 26

Permutation Matrices 27
From permutations to permutation matrices: 28
Back to the order three symmetric group: 29
The order of elements: 30
The group C_3: 30
Another view of permutations: 31
Other views of the group multiplication operation: 32
Group theory is now easy: 33
The importance of permutations: 34
Different representations: 34
Inverse permutation matrices: 35
Permutations which leave something unchanged: 36
Commutativity: 38
Abelian and non-abelian finite groups: 38
Closure of permutation matrices: 39
Summary: 39
Addendum – special relativity from permutations: 40

Contents

The Symmetric Groups 42
    Subgroups of symmetric groups: 42
    Order of the symmetric groups: 43
    Symmetric groups contain lower order symmetric groups: 44
    Summary: 45
Other Representations and Cayley Tables 46
    The adjoint representation: 49
    The Cayley table: 52
    Take care with Cayley tables: 53
    Not everything is a Cayley table: 53
    Summary: 54
An Algebraic View 56
    The order of an element: 56
    Roots of plus unity: 56
    Division Algebras from the finite groups: 57
    A word of warning: 58
    Clifford algebras reformulated: 59
The Group Axioms 60
    The inverse element: 61
    Associativity: 62
    Too many types of groups: 62
    Rotation groups: 62
    Homotopy groups and braid groups: 63
    Summary: 64
The Cyclic Groups 65
    The cyclic group permutation matrices: 65
    The cyclic groups as permutations: 67

- How many cyclic groups: 68
- Properties of cyclic groups: 68
- Cyclic groups in the complex plane: 69
- Adjoint representation of cyclic groups: 70
- Fundamental representation of the cyclic groups: 71
- Groups of prime order: 71
- Lagrange's theorem, a preview: 71
- Single groups of not prime order: 73
- The converse of Lagrange's theorem: 73
- Abelian groups and cyclic groups: 73
- Subgroups of cyclic groups: 73
- Finitely generated abelian groups: 73
- Summary: 74

The Dihedral Groups..75
- Groups and types of space: 76
- Euclidean symmetries: 77
- The dihedral groups in Euclidean space: 78
- Isomorphism between groups: 80
- The subgroup structure of the dihedral groups: 81
- Adjoint representation of the dihedral groups: 81
- Repeated warning: 84
- The Cayley table of D_4: 84
- A couple of small facts: 85
- The dihedral groups as roots of unity: 85
- Another infinite set of finite groups: 86
- The D_6 Standard Form Cayley table: 86
- Summary: 87

Contents

Group Generators 88
    Inverse elements: 89
    Cyclic groups: 89
    A group with two generators: 89
    Another example: 91
    Summary: 91

The Alternating Groups 93
    Odd permutations and even permutations: 93
    Representations of A_4 and S_4 as permutation matrices: 96
    Even permutations and odd permutations: 99
    More oddness and evenness: 99
    The adjoint representation of the symmetric groups: 101
    Summary: 101
    Addendum – determinants: 102

Subgroups 104
    Subgroups: 104
    A few theorems about subgroups: 105
    Lagrange's theorem: 105
    Cauchy's theorem: 106
    Subgroups of cyclic groups: 106
    The Sylow theorems: 106
    Little snippets: 107
    Subgroup generators: 108
    Groups of prime order: 109
    Calculating all the subgroups of a group: 109
    Summary: 110
    List of the subgroups of D_6: 110

- The Dicyclic Groups 112
  - Generating a dicyclic group: 112
  - The Standard Form Cayley table of Q_8: 113
- The Crossed Groups – Direct Products 114
  - A different type of representation: 115
  - The remarkableness of $C_2 \times C_2$ - an aside: 115
  - Crossed groups in general – the direct product: 116
  - Crossed cyclic groups: 117
  - Internal direct product: 118
  - Snippets: 118
  - Crossing permutation matrices: 118
- Finitely Generated Abelian Groups 120
  - Finitely generated groups: 120
  - Classification of finite abelian groups: 121
  - Calculation of the abelian groups of order 360: 121
  - Abelian groups of order p^n: 122
  - The Jordan-Hölder theorem: 123
- Cosets 124
  - Warning: 124
  - Cosets: 124
  - Disjoint cosets: 126
  - The index of a subgroup: 126
  - List of the cosets of D_3: 127
  - Summary: 128
- Conjugation 129
  - Finding conjugate elements within a group: 129
  - Conjugacy within abelian groups and the centre of a group: 130

Contents

- The generators of D_6: 131
- The conjugacy classes of D_6: 132
- The complete set of conjugacy classes of D_6 133
- Conjugacy technical: 133
- Permutation matrices and cycle structure: 133
- Conjugacy and cycle structure: 136
- Back to conjugacy: 136
- Geometrical interpretation of the conjugacy classes of A_4: 137
- The conjugacy classes of S_4: 138
- The number of conjugacy classes: 139
- Summary: 140

Normal Subgroups 141
- Normal subgroups: 141
- Normal subgroups of abelian groups: 142
- Other properties of normal subgroups: 142
- Normal subgroups of normal subgroups: 143
- The commutator subgroup: 144
- Notation: 145
- Summary: 145

Quotient Groups 146
- Multiplying two sets of permutations together: 146
- An example of a quotient group: 147
- Quotient groups of cyclic groups: 148

Simple Finite Groups 149
- Jordan-Hölden theorem: 149
- The classification of the finite simple groups: 149

- The simple finite groups: 150
- The 26 sporadic simple finite groups: 151
- The monster moonshine theorem: 151
- The Atlas of Finite Groups: 151

From Finite Group to Division Algebra 153
- Division algebras (types of complex numbers): 153
- Division algebras emerge from finite groups: 154
- Angles: 156
- Summary: 156

Our 4-dimensional Space-time 157
- Spinning discs: 157
- 2-dimensional rotations: 158
- From where do spaces come?: 159
- The uniqueness of our space-time: 160
- Our space-time: 162
- Electrons and neutrinos: 163
- Summary: 163

Concluding Remarks 164
Bibliography 166
Other Books by Dennis Morris 167
Index 171

Preliminaries

Preliminaries

The numbers of the finite groups to order 100

We present a list of the number of finite groups of any given order up to order one hundred. We also present a more detailed list of groups and their subgroups up to order fifteen.

Index to tables:

G is the number of groups of the given order reading across the page.

AG is the number of abelian groups of the given order reading across the page.

There is a cyclic group of every order, C_n, and so if there is only one group of a given order, say order 85, then that group is a cyclic group. The symmetric groups, S_n, are complete sets of permutations of n objects. The alternating groups, A_n, are complete sets of even permutations of n objects.

Some groups have more than one name. In these cases, the most commonly used appellation is presented before the isomorphism sign. The C stands for cyclic and C_n is the cyclic group of order n. The S stands for symmetric. For the symmetric groups, the subscript is not the order of the group. Symmetric groups, S_n, are of order $n!$. The D stands for dihedral. For the dihedral groups, the subscript is half of the order of the group. Q stands for the so called quaternion group which is the dicyclic group Q_8. For the dicyclic groups, the subscript is the order of the group. The A stands for Alternating. The alternating groups, A_n are of order $\frac{n!}{2}$. T is the cubic group which is also known as the dicyclic group Q_{12}.

Finite Groups – A Simple Introduction

The simple finite groups:
The simple finite groups (not all groups are simple) are either:

1) Cyclic groups of prime order
2) Alternating groups of order 60 or more
3) Chevalley groups
4) Twisted Chevalley groups
5) one of the 26 sporadic groups

The tables:

Order	G	AG	Order	G	AG	Order	G	AG
1	1	1	21	2	1	41	1	1
2	1	1	22	2	1	42	5	1
3	1	1	23	1	1	43	1	1
4	2	2	24	15	3	44	4	2
5	1	1	25	2	2	45	2	2
6	2	1	26	2	1	46	2	1
7	1	1	27	5	3	47	1	1
8	5	3	28	4	2	48	52	5
9	2	2	29	1	1	49	2	2
10	2	1	30	4	1	50	5	2
11	1	1	31	1	1	51	1	1
12	5	2	32	51	7	52	5	2
13	1	1	33	1	1	53	1	1
14	2	1	34	2	1	54	15	3
15	1	1	35	1	1	55	2	1
16	14	5	36	14	4	56	13	3
17	1	1	37	1	1	57	2	1
18	5	2	38	2	1	58	2	1
19	1	1	39	2	1	59	1	1
20	2	2	40	14	3	60	13	2

Preliminaries

Order	G	AG	Order	G	AG
61	1	1	81	15	5
62	2	1	82	2	1
63	4	3	83	1	1
64	267	11	84	15	2
65	1	1	85	1	1
66	4	1	86	2	1
67	1	1	87	1	1
68	5	2	88	12	3
69	1	1	89	1	1
70	4	1	90	10	2
71	1	1	91	1	1
72	50	6	92	4	2
73	1	1	93	2	1
74	2	1	94	2	1
75	3	2	95	1	1
76	4	2	96	230	7
77	1	1	97	1	1
78	6	1	98	5	2
79	1	1	99	2	2
80	52	5	100	16	4

Proper subgroups to order 15:

If a group appears alone, as C_5, then there is only one subgroup of that form within the given group. Where there are multiple subgroups of the same form, they are shown as $3(C_2)$. There are fourteen order 16 groups.

Order	Group	Proper subgroups
1 – Cyclic	C_1	-
2 – Cyclic	$C_2 \cong S_2$	-
3	C_3	-
4	C_4	C_2
4 – Crossed	$C_2 \times C_2$	$3(C_2)$

5	C_5	-
6	C_6	C_2, C_3
6 - Symmetric	$S_3 \cong D_3$	$3(C_2)$, C_3
7	C_7	-
8	C_8	C_2, C_4
8	$C_2 \times C_4$	$3(C_2)$, $2(C_4)$, $C_2 \times C_2$
8	$C_2 \times C_2 \times C_2$	$7(C_2)$, $7(C_2 \times C_2)$
8 – Dihedral	D_4	C_4, $5(C_2)$, $2(C_2 \times C_2)$
8 – Dicyclic	$Q \cong Q_8$	C_2, $3(C_4)$
9	C_9	C_3
9	$C_3 \times C_3$	$4(C_3)$
10	C_{10}	C_2, C_5
10 – Dihedral	D_5	$5(C_2)$, C_5
11	C_{11}	-
12	C_{12}	C_2, C_3, C_4, C_6
12	$C_2 \times C_6$	$3(C_2)$, C_3, $C_2 \times C_2$, $3(C_6)$
12 – Alternating	A_4	$3(C_2)$, $C_2 \times C_2$, $4(C_3)$
12	D_6	$7(C_2), C_3, C_6, 3(C_2 \times C_2), 2(D_3)$
12 – Dicyclic	$T \cong Q_{12}$	C_2, C_3, $3(C_4)$, C_6
13	C_{13}	-
14	C_{14}	C_2, C_7
14	D_7	$7(C_2)$, C_7
15	C_{15}	C_3, C_5

Introduction

Chapter 1

Introduction

Finite groups underpin much of mathematics. They are certainly the foundations of all division algebras and with that the foundation of our understanding of empty space. It has recently been realised that finite groups play at least two fundamental roles in physics; firstly, by virtue of their underpinning of division algebras, which include the Clifford algebras when they are properly formulated[1], finite groups are central to particle physics; secondly, the role finite groups play in determining the nature and uniqueness of our 4-dimensional space-time, including general relativity and classical electromagnetism, makes an understanding of finite groups essential to physicists.

Finite group theory is often seen as being dry and abstract; this is because it is usually presented in a dry and abstract way. Finite group theory is an area of mathematics in which it is very easy to drown in abstract notation and abstract proofs. Such a plethora of detail obscures the concepts and makes the subject seem boring.

Books on finite group theory are usually written by mathematicians for mathematicians, and they plod along collecting results, throwing in lemmas, proving theorems etc... This is how a technical mathematics book should be written, but to the non-mathematician, such a presentation is very dry. Actually, I am a mathematician, and I still find such a presentation very dry.

Finite groups are more than only symmetry:
Many books on finite groups have tried to introduce a geometrical aspect to finite groups to inspirit the presentation, but, in doing so, many become either childish or incarcerated within Euclidean space. We hear

---
[1] See : Dennis Morris : The Naked Spinor

people saying finite group theory is about symmetry; by this they usually mean symmetry within Euclidean space. Finite groups are above only Euclidean space, and we seek not to incarcerate them within this single one of an infinite number of different types of empty space. Finite groups are not really about symmetry; they are about permutations; it is just that regular polygons and regular polyhedra in Euclidean space have symmetries which are permutations of their apices. There is no similar 'symmetry' in other types of space such as 2-dimensional space-time.

It's the concepts that count:
Many people, including physicists, want only the results of finite group theory. These people trust the mathematicians to get it right, and they do not need to see the tedious and detailed proofs. This book is written for such people.

In this book, we describe finite groups. We avoid tedious and detailed proofs in mathematical notation, although we might outline a few proofs in less tedious prose along the way. This is not a detailed textbook on finite group theory; this is a book about finite groups. We concentrate on the concepts rather than the mathematical details. Of course, we do need some simple mathematics, such as matrix multiplication, but that is about all.

Permutation matrices:
This book is unique in that we use permutation matrices in the presentation of finite group theory. As far as your author is aware, permutation matrices have never before been used to present finite group theory. The use of permutation matrices allows the reader to 'see what is happening' far more easily than the conventional presentation of finite group theory. We feel that, by using permutation matrices, we have avoided much of the obscurity and dryness of conventional presentations of this area of mathematics and we have 'juiced up' finite group theory.

Introduction

The use of permutation matrices also gives a view of the different representations of finite groups which are so often absent from conventional presentations of finite group theory.

We hope the reader will enjoy reading this book, but we must warn the reader that there is far more to finite group theory than is presented in this book. Finite group theory is a huge area of mathematics; there is no one book, or even one set of volumes of books, which presents all finite group theory. This book aspires to be a pleasant introduction that presents the basics of finite group theory.

The purpose of this book is to provide an easy start to understanding finite group theory. We seek to lay a solid foundation upon which the reader will be able to stand when approaching more advanced texts in this area of mathematics. Clearly, there is material which we do not cover, but this material is available in a thousand other books on finite group theory which can be read and much more easily understood after reading this book.

Upon our final reading of this book, we find that we have been a little repetitive. We have said things and have then repeated these same things in a later chapter or only a few pages later. We did consider revising the text to remove these repetitions, but, after thought, we decided to leave the repetitions in the text. There is a lot to learn in finite group theory; there are a lot of facts to store within one's brain, and one cannot expect a reader to remember every fact presented to her at a first reading. Therefore, it seems sensible to remind the reader of key facts by repeating them as the reader progresses through the book. The hope is that, having been drenched in these facts and having digested this book, the reader will find more advanced texts much easier to chew.

We end the book by outlining the deduction of the nature of our 4-dimensional space-time from the finite groups. Although we skip the details, this is an eye-opening aspect of finite groups which we believe can be found in no other book on finite groups.

Chapter 2

The Permutation View of Finite Groups

A finite group is a closed set of permutations. This is called the permutation view of finite groups.

The permutation view of finite groups is the standard mantra, and, in the main, we will use the permutation view throughout most of this book. We will also briefly present an algebraic view of finite groups. The algebraic view is enlightening, but it is not well-known and it is not so easy to work with as are permutations when dealing with large finite groups. From time to time, we will also give a geometric view of finite groups, but, unlike many books, we will not rely on the geometric view. The geometric view is incarcerated within Euclidean space, and we wish to be unbound from the chains keeping us within only one kind of empty space.

Cayley's Theorem:
The fact that every finite group is a set of permutations is called Cayley's theorem. It is named after Arthur Cayley (1821-1895) who presented this theorem to the world in 1854[2].

The reader can take one of two views of Cayley's theorem. Either there are things called finite groups which we find associated with symmetries of polyhedra that 'just happen' to also be closed sets of permutations or there are things called finite groups that are closed sets of permutations that 'just happen' to be associated with symmetries of polyhedra. In this book, we take the latter view. In this book, a finite group is a closed set of permutations and all the symmetry and

---

[2] Cayley, Arthur: (1854) On the theory of groups as depending upon the symbolic equation $\theta^n = 1$: Philosophical Magazine 7 (42) 40-47 . It is interesting that Cayley sees the elements of groups as n^{th} roots of unity – see later.

polyhedra stuff is by-the-way and not of any central importance. In other words, we are studying closed sets of permutations and nothing more than closed sets of permutations.

The permutation view of finite groups:
We have the real numbers like the number 5 or the number 213. These numbers are not concrete objects; you cannot drop the number 213 on to your foot or dissolve it in your beer, but it does exist. There is a difference between shovelling 213 tons of coal and shovelling 5 tons of coal.

There is another non-concrete 'object' or concept which we call a permutation. Three objects, $\{A, B, C\}$ can be arranged, permuted, in six different ways - think of permuting three coloured balls, aubergine, blue, and cyan - in a straight line. These six different ways are:

$$\begin{array}{ccc} ABC, & BCA, & CAB \\ ACB, & BAC, & CBA \end{array} \quad (2.1)$$

You cannot drop a particular permutation on your foot or dissolve it in your beer. However, order matters. Using real numbers as our ordered objects, we see there is a difference between 123 tons of coal and 321 tons of coal.

A notation for permutations:
The standard notation for a permutation is two lines of positive whole real numbers, integers, in a bracket. Usually, but not necessarily, the top line of numbers is neatly put in ascending order starting with one. The lower line of numbers is the positions to which the numbers in the top line of numbers are moved by the permutation. Using this notation, the above permutations of coloured balls, (2.1), are:

$$ABC \equiv \begin{pmatrix} 1 & 2 & 3 \\ 1 & 2 & 3 \end{pmatrix} \quad BCA \equiv \begin{pmatrix} 1 & 2 & 3 \\ 3 & 1 & 2 \end{pmatrix} \quad CAB \equiv \begin{pmatrix} 1 & 2 & 3 \\ 2 & 3 & 1 \end{pmatrix}$$
(2.2)
$$ACB \equiv \begin{pmatrix} 1 & 2 & 3 \\ 1 & 3 & 2 \end{pmatrix} \quad BAC \equiv \begin{pmatrix} 1 & 2 & 3 \\ 2 & 1 & 3 \end{pmatrix} \quad CBA \equiv \begin{pmatrix} 1 & 2 & 3 \\ 3 & 2 & 1 \end{pmatrix}$$

This notation means:

$$BCA \equiv \begin{pmatrix} 1 & 2 & 3 \\ \downarrow & \downarrow & \downarrow \\ 2 & 3 & 1 \end{pmatrix} \equiv \begin{pmatrix} A & B & C \\ \downarrow & \downarrow & \downarrow \\ B & C & A \end{pmatrix} \qquad (2.3)$$

The permutation which changes nothing is called the identity permutation:

$$\text{Identity permutation} = \begin{pmatrix} 1 & 2 & 3 \\ 1 & 2 & 3 \end{pmatrix} \qquad (2.4)$$

The permutation notation is ambiguous. A permutation can be written in many ways because the top row of numbers does not have to be in any particular order; we have:

$$BCA \equiv \begin{pmatrix} 1 & 2 & 3 \\ 2 & 3 & 1 \end{pmatrix} = \begin{pmatrix} 1 & 3 & 2 \\ 2 & 1 & 3 \end{pmatrix} = \begin{pmatrix} 3 & 2 & 1 \\ 1 & 3 & 2 \end{pmatrix} = \begin{pmatrix} 2 & 1 & 3 \\ 3 & 2 & 1 \end{pmatrix} = \dots$$
(2.5)

We sometimes refer to a permutation of n objects as an order n permutation.

Multiplication of permutations:
In due course, we will see that we can combine two permutations together. We refer to this combination of two permutations as the multiplication of two permutations. We will present the permutation multiplication procedure later. We cannot add permutations; well; we could call the operation of combining two permutations addition rather than calling it multiplication, but we do not call it addition. The real

numbers, $\mathbb{R}$, or the complex numbers, $\mathbb{C}$, have two 'combining operations' which we call addition and multiplication. The point is that permutations have only one 'combining operation' which is arbitrarily[3] called multiplication.

An abstract view:

The reader might like to take the view that each permutation bracket, (2.2), is a symbol, like the symbol for a real number, and that these symbols, like real numbers, do not have a concrete existence. We can multiply two real numbers together even though they have no concrete existence, and we can multiply two permutations together even though they have no concrete existence. Finite group theory is the algebra of permutations. Actually, that's worth emphasizing.

Finite group theory is the algebra of permutations.

We do want the reader to get this idea settled solidly into her understanding, and so we will repeat it once more.

Finite group theory is the algebra of permutations. Instead of doing algebra with numbers, we are doing algebra with permutations.

The algebra of permutations has a multiplication operation, but it does not have an addition operation.

Mappings:

A technically correct but rather obscure and boring way to view a permutation is as a mapping from n real integers to n real integers. If you like this abstract view, then we are glad to have presented it to you. If you do not like this abstract view, then you will lose nothing by forgetting all about it.

---

[3] When we look at permutation matrices, we see that the 'combining operation' is actually matrix multiplication, and so the use of the word 'multiplication' is not entirely arbitrary.

Back to the permutation view of finite groups:
Above, (2.1) we have all six possible permutations of three objects. We see that we can permute three objects in six and in only six different ways. The whole six permutations are a complete set of permutations. They are the permutations of three objects. As we formed six permutations of three objects above, (2.1), so we can form 24 permutations of four objects or 120 permutations of five objects, and so on. In each case, we have a definite number of permutations – 6, 24, 120…

However we combine, multiply together, the six permutations above, (2.2), we will get a permutation which is one of this set of six permutations. This has to be the case because there are only these six permutations. We say this set of six permutations is closed; it is a closed set of permutations. Although it is not yet clear to the reader because we have not yet explained the permutation multiplication procedure, the three permutations in the top row of (2.2) are also a closed set of permutations.

A closed set of permutations is a finite group, and a finite group is a closed set of permutations.

Actually, this too is worth repeating. A finite group is nothing more than a closed set of permutations together with 'permutation multiplication'. Nothing else is a finite group. We say this once more.

A closed set of permutations is a finite group, and a finite group is a closed set of permutations.

Finite Groups:
The permutation view of finite groups is that a finite group is a closed set of permutations. For example, the six permutations above, (2.1), are the finite group we call the symmetric group S_3; the S stands for symmetric, and the 3 is the number of objects permuted. The three permutations in the top row of (2.2) are the finite group we denote as C_3; the C is for cyclic. The symmetric group S_4 is the 24 permutations of four objects.

Note that the finite group is not the objects being permuted although even specialist group theorists often secretly think of the objects this way. The finite group is the permutations. We do not think of one ice cream as being the number one or of two pints of beer as being the number two, and so it is with finite groups; the ordered objects, $\{A, B, C\}$ are not the finite group just as the two pints of beer are not the number two.

So, when we have a way of multiplying permutations together, if we have a set of permutations such that every possible product of two or more of these permutations is another member of the set, then this set of permutations is a finite group. It is a finite group because it is closed under permutation multiplication.

The group order and why finite:
Each closed set of permutations is a finite number of permutations. The number of the permutations in a finite group is called the 'order' of the group. This word 'order' should not be confused with the order of the objects in a particular permutation. The order of a finite group is simply the size of that finite group – how many permutations there are in that finite group. There are six permutations in the symmetric group S_3, and so the order of S_3 is six. The order of S_4 is 24. The order of some groups is very large; like the real numbers, the orders of the finite groups march all the way to infinity. The order of the symmetric group of all permutations of n objects is n-factorial[4], $n!$, and n can be any whole positive real number. There are infinitely many symmetric finite groups of orders $2!, 3!, 4!, \ldots n! \ldots \infty$. The finite groups are called finite because each finite group is a finite number of permutations.

The order of a finite group, G, is often written as $|G|$; for example, $|S_3| = 6$.

---

[4] $n! = n \times (n-1) \times (n-2) \times \ldots \times 3 \times 2 \times 1$. Zero factorial is defined to be 1.

Finite Groups – A Simple Introduction

The word finite is used to distinguish these groups of permutations from the infinite rotation groups. A rotation group is a spherical surface, like a circle or a sphere, which has infinitely many points upon it. Rotation groups[5] are often called infinite groups, and it is unfortunate, though understandable,[6] that these two very different mathematical objects should share the name 'group'. In this book, we have no interest in the, so called, infinite groups.

Squares and triangles:

The reader might already have been told that a finite group is something to do with regular polygons like squares or triangles. Finite group theory is often taught this way. For example, the three corners of a triangle can be permuted in six ways (see drawing below). If the triangle is equilateral, this is equivalent to rotating the triangle or turning the triangle over.

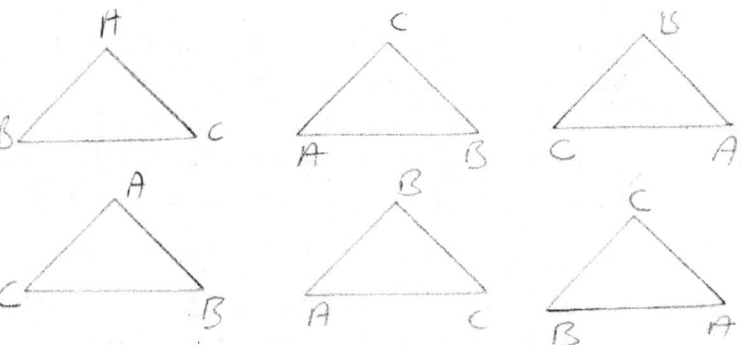

The reader might notice that the top row of triangles are rotations without flipping over the triangle; these are a closed sub-set of the six apex permutations in the drawing; this is the cyclic finite group of order three denoted by C_3. The complete set of six apex permutations is the symmetric finite group of order six denoted by S_3. We see that S_3 contains a C_3 subgroup.

---

[5] The rotation groups are considered in : Dennis Morris : Lie Groups and Lie Algebras.
[6] This coincidence of names is due to both entities satisfying the same set of axioms; the axioms are not choosy enough.

The four corners of a square can be permuted in 24 ways. We see that this 'triangle stuff' is just permutations.

Your author does not like this triangle/square way of presenting a finite group although there are some pedagogic advantages to such a presentation. Triangles, squares, and polygons in general exist in only 2-dimensional Euclidean space, $\mathbb{E}^2$. There are no polygons in 2-dimensional space-time because we cannot rotate through 360^0 within 2-dimensional space-time. Rotation in 2-dimensional space-time is change of velocity, and, sooner or later, we hit the light barrier. Presenting finite groups using polygons ties us to Euclidean space whereas a special set of permutations is, like the real numbers, something over and above a particular type of space.

Similarly, the apices of a Euclidean regular polyhedron can be permuted, and the complete set of permutations is a finite group. Of course, Euclidean regular polyhedra exist in only 3-dimensional Euclidean space, $\mathbb{E}^3$.[7]

We will return to Euclidean space from time to time because there are some pedagogic advantages in using it to demonstrate finite groups, but it is not of the essence of finite group theory.

Summary so far:
The standard mantra is that a finite group is a closed set of permutations. Permutations can be combined, multiplied, together. The finite group of permutations is closed under this permutation multiplication operation. The finite group of permutations includes the identity permutation.

A permutation is a non-concrete concept symbolised by, for examples:

---

[7] Mathematicians have imagined higher dimensional forms of Euclidean space. Within these spaces, there exist higher dimensional 'polyhedra' which are called polytopes. There is a whole area of mathematics concerning polytopes.

$$\begin{pmatrix} 1 & 2 & 3 \\ 3 & 2 & 1 \end{pmatrix} \quad \text{or} \quad \begin{pmatrix} 1 & 3 & 4 & 2 \\ 4 & 2 & 1 & 3 \end{pmatrix} = \begin{pmatrix} 1 & 2 & 3 & 4 \\ 4 & 3 & 2 & 1 \end{pmatrix} \qquad (2.6)$$

We have not yet given a complete overview of finite groups; there is more to come, but this chapter is already too long. We seek to present this subject in bite size chunks, and so we will end this initial chapter here.

Time for a cup of tea.

Products of Permutations

Chapter 3

Products of Permutations

What is a permutation?:
A permutation is an action which starts with a set of objects in a particular order and swaps that order about. For clarity, we will consider three objects initially in the order ABC, and we will swap the order. We have the six permutations:

$$\begin{array}{lll} ABC \to ABC & ABC \to BCA & ABC \to CAB \\ ABC \to BAC & ABC \to CBA & ABC \to ACB \end{array} \quad (3.1)$$

Now comes the clever bit; we can combine permutations. For example, we begin with the permutation:

$$ABC \to BCA \equiv \begin{pmatrix} 1 & 2 & 3 \\ 2 & 3 & 1 \end{pmatrix} \quad (3.2)$$

This permutation, (3.2), says take the object in the first place and put it in the second place, take the object in the second place and put it in the third place, and take the object in the third place and put it in the first place.

We now permute the RHS of (3.2) again. We begin this second permutation with B in the first place, C in the second place, and A in the third place and we apply the permutation:

$$ABC \to CAB \equiv \begin{pmatrix} 1 & 2 & 3 \\ 3 & 1 & 2 \end{pmatrix} \quad (3.3)$$

This permutation, (3.3), says take the object in the first place and put it in the third place, take the object in the second place and put it in the first place, and take the object in the third place and put it in the second place.

Acting upon BCA, (3.2), with this second permutation, (3.3), gives:

$$ABC \to BCA \to ABC \equiv \begin{pmatrix} 1 & 2 & 3 \\ 2 & 3 & 1 \end{pmatrix}\begin{pmatrix} 1 & 2 & 3 \\ 3 & 1 & 2 \end{pmatrix} = \begin{pmatrix} 1 & 2 & 3 \\ 1 & 2 & 3 \end{pmatrix} \quad (3.4)$$

You do not have to be a genius to realise that, if you combine two permutations together in this way, you will get another permutation. In this case, (3.4), the resultant permutation is the identity permutation – change nothing.

Let's go through that again. We combine permutations by first acting with one permutation upon three objects and then acting with a second permutation upon the three previously permuted objects. In every case, the result will be as if we had acted with a single permutation. This is like first multiplying by 3 and then multiplying by 2 is the same as multiplying by 6.

This sequential combination of permutations is called permutation multiplication; remarkably, it is matrix multiplication in disguise.

Closure again:
If we combine two permutations together, we get another permutation. No matter how we combine the six permutations of three objects given above, (3.1), we will always get one of the six permutations of three objects – there are no other permutations to get. This is called closure. Closure is quite obvious; no matter how we combine the six permutations, we do not get another different permutation – there are no different permutations to get.

The set of six permutations, (3.1), is said to be closed under combination of permutations. Of course, the same is true of the 24 permutations of four objects and of the 120 permutations of five objects, and so on.

Imagine that we chose only four of the six permutations above, (3.1), and we began combining these four permutations together in various ways; it would not be long before we produced both of the two unchosen permutations giving us the whole set of six permutations. The set of four permutations is not a closed set of permutations. The property of closure is an essential property of a finite group. Hence the set of four permutations is not a finite group.

Closure is usually stated as 'every combination of two elements of the finite group is an element of the finite group'.

The astute reader might have noticed that if we chose only the top three permutations of, (3.1), or even only one[8] of these three:

$$ABC \to ABC \qquad ABC \to BCA \qquad ABC \to CAB \qquad (3.5)$$

then these three permutations are also closed. No combination of these three permutations will produce one of the missing three permutations. These three permutations are a finite group in their own right. We met these three permutations above when we looked at triangles. These three permutations are the order three cyclic group denoted as C_3. We again see that the symmetric group S_3, which is all six of the above permutations, (3.1), has a C_3 subgroup, (3.5). The essence of this subgroup is that the order ABC stays the same but the permutation starts at a different point.

Of course, the single identity permutation is closed – no matter how many times you multiply it by itself, you will get the identity permutation. Thus, the identity permutation by itself is a finite group. We generally denote this single element finite group as C_1.

Lots of mathematical 'things' are closed. If we multiply two real numbers together, we get a real number. If we multiply two complex numbers together, we get a complex number. Two rotation matrices of the same form multiplied together produce a rotation matrix of the same form. We might refer to closure as 'a duck mated with a duck produces a duck' – not a chicken. There are mathematical objects which are not closed; two vectors dotted together produce a real number and not a vector. This is 'a duck mated with a duck producing a robin'. Above, we have an order 3 permutation mated with an order 3 permutation producing an order 3 permutation.

---

[8] Other than the identity permutation.

Identity:

We see that one of the permutations in S_3, (3.1), changes nothing:

$$ABC \to ABC \qquad (3.6)$$

We call this the identity permutation. We see that this permutation is also one of the three permutations which are the finite group C_3.

Every finite group has the identity permutation within it.

Within the literature, the identity permutation, often called the identity element of the group or just the identity, is usually denoted by the lower case letter e. Although we will use the conventional lower case letter e to denote the identity of a finite group from time to time, we will also use the lower case letter a to denote the identity when we look at permutation matrices.

Powers of permutations:

If you choose any one of the five non-identity S_3 permutations above, (3.1), and you successively combine this permutation with itself enough times, you will eventually get the identity permutation. We use the superscript to denote the power to which the permutation is raised; we have:

$$
\begin{array}{lll}
ABC^1 \to ABC & BCA^3 \to ABC & CAB^3 \to ABC \\
BAC^2 \to ABC & CBA^2 \to ABC & ACB^2 \to ABC
\end{array} \qquad (3.7)
$$

$$\begin{pmatrix} 1 & 2 & 3 \\ 1 & 2 & 3 \end{pmatrix}^1 = \begin{pmatrix} 1 & 2 & 3 \\ 1 & 2 & 3 \end{pmatrix} \qquad \begin{pmatrix} 1 & 2 & 3 \\ 2 & 3 & 1 \end{pmatrix}^3 = \begin{pmatrix} 1 & 2 & 3 \\ 1 & 2 & 3 \end{pmatrix} \qquad (3.8)$$

$$\begin{pmatrix} 1 & 2 & 3 \\ 3 & 1 & 2 \end{pmatrix}^3 = \begin{pmatrix} 1 & 2 & 3 \\ 1 & 2 & 3 \end{pmatrix} \qquad \begin{pmatrix} 1 & 2 & 3 \\ 1 & 3 & 2 \end{pmatrix}^2 = \begin{pmatrix} 1 & 2 & 3 \\ 1 & 2 & 3 \end{pmatrix} \qquad (3.9)$$

$$\begin{pmatrix} 1 & 2 & 3 \\ 2 & 1 & 3 \end{pmatrix}^2 = \begin{pmatrix} 1 & 2 & 3 \\ 1 & 2 & 3 \end{pmatrix} \qquad \begin{pmatrix} 1 & 2 & 3 \\ 3 & 2 & 1 \end{pmatrix}^2 = \begin{pmatrix} 1 & 2 & 3 \\ 1 & 2 & 3 \end{pmatrix} \qquad (3.10)$$

Products of Permutations

Since a finite group is a closed set of permutations, and any permutation combined with itself enough times will produce the identity permutation, the finite group must include the identity permutation.

This is an important point; let us re-emphasize it. Every permutation, when multiplied by itself enough times, will eventually become the identity permutation. Every element of a finite group, when raised to the appropriate power, will become the identity element of the finite group.

The order of a permutation:
We say that the power to which a permutation is raised to make it equal to the identity is the order of that permutation. This is the order of a single permutation; this is not the order, size, of the finite group containing this permutation, and it is not the order of the coloured balls. There is perhaps too much use of the word order in many different ways in finite group theory.

Since every element of a finite group, G, has a particular finite order, we can see a finite group as nothing more than a set of elements each of which is a nth root of the identity but which are 'tangled' together. Perhaps the reader should read the previous sentence again.

To jump the gun, the order of an element of a finite group is less than or equal to the order of the finite group, $|G|$. Furthermore, the order of each and every element of a finite group divides the order of the finite group. We will look at this in more detail later.

Non-commutativity of permutations:
Permutations, are non-commutative, there are two products of two permutations depending upon the order of the multiplication.

Take the two permutations:

$$\begin{pmatrix} 1 & 2 & 3 & 4 & 5 \\ 3 & 5 & 4 & 1 & 2 \end{pmatrix} \begin{pmatrix} 1 & 2 & 3 & 4 & 5 \\ 3 & 5 & 1 & 2 & 4 \end{pmatrix} \quad (3.11)$$

We are going to form the product of these two permutations. We could choose to 'multiply' left to right, but we could equally well choose to 'multiply' right to left. We will get two different answers.

With any matrices, we could similarly choose to 'multiply' left to right or we could equally well choose to 'multiply' right to left. Mathematically, it matters not which way we choose to multiply matrices, but, having chosen one way, we must stick with that choice. With matrices, the choice was made 150 years ago to multiply from left to right.

We will see in the next chapter that permutations are very closely connected to matrices. With permutations, the choice was made to multiply from right to left to fit with the choice made to multiply matrices from left to right[9]. We could multiply permutations left to right and still fit with matrix multiplication, but we would have to change lots of the permutation notation, and so we will not bother. Just remember: matrices left to right; permutations right to left.

The multiplication of permutations:
We multiply the two permutations, (3.11), by starting with the rightmost permutation. Remember right to left. We begin at the number 1 in the top row of the rightmost permutation. The number 1 goes to the number 3 in the bottom row of the rightmost permutation. We now go to the number 3 in the top row of the leftmost permutation in (3.11). This number 3 goes to the number 4 in the bottom row of the leftmost permutation. Thus we have the first part of our product:

$$1 \to 3, 3 \to 4 \Rightarrow \begin{pmatrix} 1 \\ 4 \end{pmatrix} \qquad (3.12)$$

Similarly:

---

[9] Actually, the correspondence between matrix multiplication and permutation multiplication seems to have arisen 'unconsciously' within mathematics. You author can find no statement of this correspondence, which will be presented in the following chapter, anywhere in the mathematical literature prior to this book.

Products of Permutations

$$2 \to 5, 5 \to 2: \quad 3 \to 1, 1 \to 3: \quad 4 \to 2, 2 \to 5: \quad 5 \to 4, 4 \to 1$$

$$\begin{pmatrix} 1 & 2 & 3 & 4 & 5 \\ 3 & 5 & 4 & 1 & 2 \end{pmatrix} \begin{pmatrix} 1 & 2 & 3 & 4 & 5 \\ 3 & 5 & 1 & 2 & 4 \end{pmatrix} = \begin{pmatrix} 1 & 2 & 3 & 4 & 5 \\ 4 & 2 & 3 & 5 & 1 \end{pmatrix}$$
(3.13)

We point out that (3.13) is matrix multiplication (in the case of permutation matrices) in an obscure but comparatively easily handled notation. This will become clearer in the next chapter. Hm! it does not look like matrix multiplication, does it?

Another example:
Well, it is the same example, but with the order reversed. We have:

$$1 \to 3, 3 \to 1: \quad 2 \to 5, 5 \to 4: \quad 3 \to 4, 4 \to 2$$
$$4 \to 1, 1 \to 3: \quad 5 \to 2, 2 \to 5$$

(3.14)

$$\begin{pmatrix} 1 & 2 & 3 & 4 & 5 \\ 3 & 5 & 1 & 2 & 4 \end{pmatrix} \begin{pmatrix} 1 & 2 & 3 & 4 & 5 \\ 3 & 5 & 4 & 1 & 2 \end{pmatrix} = \begin{pmatrix} 1 & 2 & 3 & 4 & 5 \\ 1 & 4 & 2 & 3 & 5 \end{pmatrix}$$

Comparing (3.13) and (3.14), we note that permutation multiplication is non-commutative, as we said above.

We also note that in the first multiplication, (3.13), two positions, $\{2,3\}$ were unchanged and that in the second multiplication, (3.14), we still have two positions unchanged, $\{1,5\}$.

Warning:
In case the reader has not previously met non-commutative multiplication, we point out that in finite group theory we cannot simply do multiplication as we are accustomed doing multiplication within the real numbers. We must be careful to keep the order of the elements unchanged. For example, if $\{d, g, f\}$ are three different elements of a

finite group, we cannot simply write their product as these three letters in any order. Indeed, we have six different products of these three elements. When forming a product, we have to specify whether we are multiplying on the left or multiplying on the right. We give some examples.

Multiplying d on the left by g is the product gd whereas multiplying d on the right by g is the product dg. In general, these two products are not equal. We can then multiply the product dg on the right by g to give dg^2 or on the left by g to give gdg. In general, these two products will not be the same. We must be careful to keep the elements in order.

Of course, any element of a group always commutes with itself; dd is the same product as dd. Any element of a group also always commutes with its own inverse, but more on that soon.

Inverse permutations:

The inverse of a permutation, like the inverse of any other mathematical object, is the permutation which when multiplied by the original permutation gives the identity permutation.

We form the inverse of a permutation by simply swapping the two rows of the permutation. We have:

$$\begin{pmatrix} 1 & 2 & 3 & 4 \\ 3 & 2 & 4 & 1 \end{pmatrix}^{-1} = \begin{pmatrix} 3 & 2 & 4 & 1 \\ 1 & 2 & 3 & 4 \end{pmatrix} \qquad (3.15)$$

We have:

$$\begin{pmatrix} 1 & 2 & 3 & 4 \\ 3 & 2 & 4 & 1 \end{pmatrix} \begin{pmatrix} 3 & 2 & 4 & 1 \\ 1 & 2 & 3 & 4 \end{pmatrix} = \begin{pmatrix} 1 & 2 & 3 & 4 \\ 1 & 2 & 3 & 4 \end{pmatrix} \qquad (3.16)$$

This is the same as:

$$\begin{pmatrix} 1 & 2 & 3 & 4 \\ 3 & 2 & 4 & 1 \end{pmatrix} \begin{pmatrix} 1 & 2 & 3 & 4 \\ 4 & 2 & 1 & 3 \end{pmatrix} = \begin{pmatrix} 1 & 2 & 3 & 4 \\ 1 & 2 & 3 & 4 \end{pmatrix} \qquad (3.17)$$

Products of Permutations

wherein we have just made the inverse a little neater. We also have:

$$\begin{pmatrix} 1 & 2 & 3 & 4 \\ 4 & 2 & 1 & 3 \end{pmatrix} \begin{pmatrix} 1 & 2 & 3 & 4 \\ 3 & 2 & 4 & 1 \end{pmatrix} = \begin{pmatrix} 1 & 2 & 3 & 4 \\ 1 & 2 & 3 & 4 \end{pmatrix} \qquad (3.18)$$

We see that a permutation commutes with its inverse. Of course it does, both products are the identity; both products are the same.

We see that a permutation has only one inverse; it can be written in different ways, but there is still only one inverse of any permutation.

We form the inverse of a permutation by 'flipping' the permutation. By this we mean that we swap the top row for the bottom row and we swap the bottom row for the top row. Easy peasy. We can then tidy the notation if we wish.

Cycle structure and a note on notation:
We sometime see a permutation split apart; for example:

$$\begin{pmatrix} 1 & 2 & 3 & 4 & 5 & 6 \\ 1 & 6 & 5 & 6 & 4 & 2 \end{pmatrix} \equiv \begin{pmatrix} 1 \\ 1 \end{pmatrix} \begin{pmatrix} 2 & 6 \\ 6 & 2 \end{pmatrix} \begin{pmatrix} 3 & 4 & 5 \\ 5 & 6 & 4 \end{pmatrix} \qquad (3.19)$$

We can do this because the separate parts of the permutation do not mix together. We say the separate parts of the permutation are disjoint.

Often, the single element brackets are omitted; for example:

$$\begin{pmatrix} 1 & 2 & 3 & 4 & 5 & 6 \\ 1 & 6 & 3 & 5 & 4 & 2 \end{pmatrix} \equiv \begin{pmatrix} 1 \\ 1 \end{pmatrix} \begin{pmatrix} 3 \\ 3 \end{pmatrix} \begin{pmatrix} 2 & 6 \\ 6 & 2 \end{pmatrix} \begin{pmatrix} 4 & 5 \\ 5 & 4 \end{pmatrix}$$

$$\equiv \begin{pmatrix} 2 & 6 \\ 6 & 2 \end{pmatrix} \begin{pmatrix} 4 & 5 \\ 5 & 4 \end{pmatrix} \qquad (3.20)$$

$$\equiv \begin{pmatrix} 2 & 4 & 5 & 6 \\ 6 & 5 & 4 & 2 \end{pmatrix}$$

We do not like this omission of single cycles, but that is no more than personal taste. We will not omit single cycles in this book.

The disjoint notation, with single elements, is called the cycle structure of the permutation. The cycle structure is useful notation, but it need not be used when, for example, multiplying permutations. You might sometimes be asked to multiply together two permutations which are in disjoint form with single elements missing. My advice is to write the two permutations as full permutations and then multiply them together.

The cycle structure of the permutation holds a lot of useful information about the permutation. When we look at conjugate elements within a group, we will see that elements which are conjugate to each other have the same cycle structure and are of the same order as each other.

By the way, the separate cycles commute with each other.

The order of a permutation from the cycle structure:
The order of the permutation, group element, above, (3.19), is six. This is the lowest common multiple of the lengths of the different disjoint cycles. The order of the permutation:

$$\begin{pmatrix} 1 & 2 & 3 & 4 & 5 & 6 & 7 & 8 & 9 \\ 3 & 1 & 2 & 9 & 4 & 7 & 8 & 6 & 5 \end{pmatrix} = \begin{pmatrix} 1 & 2 & 3 \\ 3 & 1 & 2 \end{pmatrix} \begin{pmatrix} 4 & 5 & 9 \\ 9 & 4 & 5 \end{pmatrix} \begin{pmatrix} 6 & 7 & 8 \\ 7 & 8 & 6 \end{pmatrix}$$
(3.21)

is three because three is the lowest common multiple of the three cycles of length three. When we look at permutation matrices, this will be much easier to understand.

Summary:
Permutation multiplication is just sequentially combining permutations. We multiply permutations from right to left. Permutation multiplication is non-commutative.

The inverse permutation is formed by swapping the two rows.

We can write a permutation in cycle structure format.

Chapter 4

Permutation Matrices

There is a one-to-one correspondence between permutations and a particular type of square matrices called permutation matrices. We will use the correlation to assist us in expounding finite group theory.

A permutation matrix is a square matrix with a single one in each row and a single one in each column and with zeros everywhere else. There is only one 1×1 permutation matrix; it is:

$$[1] \tag{4.1}$$

There is only one way to permute a single object.

There are only two 2×2 permutation matrices just as there are only two permutations of two objects, $AB \& BA$; these two 2×2 permutation matrices are:

$$\begin{bmatrix} 1 & 0 \\ 0 & 1 \end{bmatrix} \quad \& \quad \begin{bmatrix} 0 & 1 \\ 1 & 0 \end{bmatrix} \tag{4.2}$$

There are six 3×3 permutation matrices. We will correlate these with the six permutations of three objects given above, (2.1). We have:

$$\begin{pmatrix} 1 & 2 & 3 \\ 1 & 2 & 3 \end{pmatrix} \equiv \begin{bmatrix} 1 & 0 & 0 \\ 0 & 1 & 0 \\ 0 & 0 & 1 \end{bmatrix} \quad \begin{pmatrix} 1 & 2 & 3 \\ 3 & 1 & 2 \end{pmatrix} \equiv \begin{bmatrix} 0 & 1 & 0 \\ 0 & 0 & 1 \\ 1 & 0 & 0 \end{bmatrix}$$

$$\begin{pmatrix} 1 & 2 & 3 \\ 2 & 3 & 1 \end{pmatrix} \equiv \begin{bmatrix} 0 & 0 & 1 \\ 1 & 0 & 0 \\ 0 & 1 & 0 \end{bmatrix} \tag{4.3}$$

$$\begin{pmatrix} 1 & 2 & 3 \\ 1 & 3 & 2 \end{pmatrix} \equiv \begin{bmatrix} 1 & 0 & 0 \\ 0 & 0 & 1 \\ 0 & 1 & 0 \end{bmatrix} \qquad \begin{pmatrix} 1 & 2 & 3 \\ 2 & 1 & 3 \end{pmatrix} \equiv \begin{bmatrix} 0 & 1 & 0 \\ 1 & 0 & 0 \\ 0 & 0 & 1 \end{bmatrix}$$

$$\begin{pmatrix} 1 & 2 & 3 \\ 3 & 2 & 1 \end{pmatrix} \equiv \begin{bmatrix} 0 & 0 & 1 \\ 0 & 1 & 0 \\ 1 & 0 & 0 \end{bmatrix} \tag{4.4}$$

From permutations to permutation matrices:

Looking at the above, (4.3) & (4.4), we can see that the permutations can be mapped directly to the permutation matrices by the simple mechanism of identifying the top row of the permutation with the column of the permutation matrix and the bottom row of the permutation with the row in the permutation matrix and placing 1's in these positions:

$$\begin{pmatrix} col1 & col2 & col3 \\ row3 & row2 & row1 \end{pmatrix}$$
$$\downarrow$$
$$\begin{bmatrix} 0 & 0 & 1 \\ 0 & 1 & 0 \\ 1 & 0 & 0 \end{bmatrix} \tag{4.5}$$

This is general; it works for all permutations. The identity permutation is the identity matrix:

$$\begin{pmatrix} 1 & 2 & 3 \\ 1 & 2 & 3 \end{pmatrix}$$
$$\downarrow$$
$$\begin{bmatrix} 1 & 0 & 0 \\ 0 & 1 & 0 \\ 0 & 0 & 1 \end{bmatrix} \tag{4.6}$$

Permutation Matrices

Back to the order three symmetric group:

The finite group S_3 is the six permutations of three objects. We have three kinds of permutations in this group, S_3. In S_3, we have a single identity permutation which is equal to the identity when raised to power one; we have:

$$\begin{bmatrix} 1 & 0 & 0 \\ 0 & 1 & 0 \\ 0 & 0 & 1 \end{bmatrix}^1 = \begin{bmatrix} 1 & 0 & 0 \\ 0 & 1 & 0 \\ 0 & 0 & 1 \end{bmatrix} \tag{4.7}$$

In S_3, we have two permutations which need to be raised to power three to equal the identity; we have:

$$\begin{bmatrix} 0 & 1 & 0 \\ 0 & 0 & 1 \\ 1 & 0 & 0 \end{bmatrix}^3 = \begin{bmatrix} 1 & 0 & 0 \\ 0 & 1 & 0 \\ 0 & 0 & 1 \end{bmatrix} \quad \begin{bmatrix} 0 & 0 & 1 \\ 1 & 0 & 0 \\ 0 & 1 & 0 \end{bmatrix}^3 = \begin{bmatrix} 1 & 0 & 0 \\ 0 & 1 & 0 \\ 0 & 0 & 1 \end{bmatrix} \tag{4.8}$$

In S_3, we have three permutations which need to be raised to power two to equal the identity; we have:

$$\begin{bmatrix} 1 & 0 & 0 \\ 0 & 0 & 1 \\ 0 & 1 & 0 \end{bmatrix}^2 = \begin{bmatrix} 1 & 0 & 0 \\ 0 & 1 & 0 \\ 0 & 0 & 1 \end{bmatrix} \quad \begin{bmatrix} 0 & 1 & 0 \\ 1 & 0 & 0 \\ 0 & 0 & 1 \end{bmatrix}^2 = \begin{bmatrix} 1 & 0 & 0 \\ 0 & 1 & 0 \\ 0 & 0 & 1 \end{bmatrix}$$

$$\begin{bmatrix} 0 & 0 & 1 \\ 0 & 1 & 0 \\ 1 & 0 & 0 \end{bmatrix}^2 = \begin{bmatrix} 1 & 0 & 0 \\ 0 & 1 & 0 \\ 0 & 0 & 1 \end{bmatrix} \tag{4.9}$$

We might say that we have a group which has one identity, two cube roots of that identity and three square roots of that identity[10].

The reader might like to take a pause and examine the six permutation matrices above, (4.7) to (4.9). We see that they are of three different

---

[10] Remember Cayley (first footnote of chapter 2) and his $\theta^n = 1$ view of permutations.

types and that these different types can be associated with the number of 1's on the leading diagonal[11] of the matrices.

The order of elements:
We reiterate a little in this next paragraph. The order of a finite group is the size of that finite group; this is the number of elements in that finite group. Unfortunately, we also use the word 'order' in relation to each element of a group.

Every element in a finite group will, if raised to the appropriate non-zero power, be equal to the identity. We have seen an example of this above, (4.7) to (4.9), in the group S_3. The power to which an element is raised is called the order of that element. Every element of every finite group has an order associated with it; for example, we say that the elements in (4.9) are of order two.

The group C_3:
The identity and the two cube roots of the identity are the order three cyclic group, C_3; we have:

$$C_3 = \left\{ \begin{bmatrix} 1 & 0 & 0 \\ 0 & 1 & 0 \\ 0 & 0 & 1 \end{bmatrix}, \begin{bmatrix} 0 & 1 & 0 \\ 0 & 0 & 1 \\ 1 & 0 & 0 \end{bmatrix}, \begin{bmatrix} 0 & 0 & 1 \\ 1 & 0 & 0 \\ 0 & 1 & 0 \end{bmatrix} \right\} \qquad (4.10)$$

In the earlier chapter, circa (3.13), we combined, multiplied, permutations together from the right to the left. We combine permutation matrices together simply by matrix multiplication. Of course, matrix multiplication is from the left to the right. For example:

---

[11] The leading diagonal is from top left-hand corner to bottom right-hand corner.

Permutation Matrices

$$\begin{bmatrix} 1 & 0 & 0 \\ 0 & 0 & 1 \\ 0 & 1 & 0 \end{bmatrix} \begin{bmatrix} 0 & 0 & 1 \\ 0 & 1 & 0 \\ 1 & 0 & 0 \end{bmatrix} = \begin{bmatrix} 0 & 0 & 1 \\ 1 & 0 & 0 \\ 0 & 1 & 0 \end{bmatrix}$$
$$\downarrow \qquad\qquad \downarrow \qquad\qquad \downarrow \qquad\qquad (4.11)$$
$$\begin{pmatrix} 1 & 2 & 3 \\ 1 & 3 & 2 \end{pmatrix} \begin{pmatrix} 1 & 2 & 3 \\ 3 & 2 & 1 \end{pmatrix} = \begin{pmatrix} 1 & 2 & 3 \\ 2 & 3 & 1 \end{pmatrix}$$

We have vertically correlated the permutations and the permutation matrices. We multiplied the matrices from left to right, as is the convention; to match this matrix convention, we have to multiply the permutations from right to left. Now you know why permutations are multiplied from right to left and why we call the combination of two permutations multiplication. It is sobering to realise that acting upon a number of objects by permuting them is matrix multiplication. We see a deep connection here between permutations, which are the stuff of finite groups, and matrix multiplication.

Another view of permutations:
We can take the view that every permutation is a permutation matrix and the permutation multiplication operation is matrix multiplication in a strange notation. Since a finite group is a closed set of permutations, we can take the view that a finite group is a closed set of permutation matrices with matrix multiplication as the group operation; this is an entirely sensible view.

Aside: In spite of mathematicians thinking mathematics to be entirely logical, there is much subjectivity in the subject. Most group theorists take the subjective view that groups are centrally concerned with symmetry because of the rotating polyhedra in Euclidean spaces. Your author takes the subjective view that Euclidean symmetries are 'small potatoes' on the periphery of group theory. Finite group theory is about permutations which are really matrices; Hm!, perhaps matrices are really permutations. Both views are subjective, and both views are valid. Neither view is logically derived, but, clearly, your author's view is the correct view.

Other views of the group multiplication operation:

The standard mantra is that although a finite group is a concept over and above the real concrete world, different groups of different objects have different operations as the group multiplication operation. For example, the three rotations of an equilateral triangle form a group in which the group operation is rotation through 120^0. The orders of three coloured balls in a straight line form a group in which the group operation is swapping balls about. Of course, the permutation group operation is permutation multiplication.

All these operations are really matrix multiplication in disguise; technically, all group operations are linear. Every finite group is a set of permutations, and permutation multiplication is just matrix multiplication.

Let us consider the rotations of an equilateral triangle. The three apices of the triangle are at the points:

$$(1,0), \left(-\frac{1}{2}, \frac{\sqrt{3}}{2}\right), \left(-\frac{1}{2}, -\frac{\sqrt{3}}{2}\right) \qquad (4.12)$$

The triangle 'sits' in 2-dimensional Euclidean space. 2-dimensional Euclidean space is the complex plane. Writing the three points above, (4.12), as complex numbers (in matrix notation),

$$\mathbb{C} \equiv a + ib \equiv \begin{bmatrix} a & b \\ -b & a \end{bmatrix} \qquad (4.13)$$

We have:

$$(1,0) \equiv \begin{bmatrix} 1 & 0 \\ 0 & 1 \end{bmatrix}$$

$$\left(-\frac{1}{2}, \frac{\sqrt{3}}{2}\right) \equiv \begin{bmatrix} -\frac{1}{2} & \frac{\sqrt{3}}{2} \\ -\frac{\sqrt{3}}{2} & -\frac{1}{2} \end{bmatrix}, \quad \left(-\frac{1}{2}, -\frac{\sqrt{3}}{2}\right) \equiv \begin{bmatrix} -\frac{1}{2} & -\frac{\sqrt{3}}{2} \\ \frac{\sqrt{3}}{2} & -\frac{1}{2} \end{bmatrix} \qquad (4.14)$$

Permutation Matrices

These three 2×2 matrices, (4.14), are a 2×2 matrix representation of the order 3 finite group C_3. The 120^0 rotation operation is just permutation matrix multiplication written in 2×2 matrices rather than in 3×3 matrices.

Of course, the multiplication operation of the complex numbers, $\mathbb{C}$, is just matrix multiplication[12] as is the case with all types of complex numbers.

We could have written the above three complex numbers as the three complex numbers in polar form:

$$\begin{bmatrix} \cos(0) & \sin(0) \\ -\sin(0) & \cos(0) \end{bmatrix}$$
$$\begin{bmatrix} \cos(120) & \sin(120) \\ -\sin(120) & \cos(120) \end{bmatrix}, \begin{bmatrix} \cos(240) & \sin(240) \\ -\sin(240) & \cos(240) \end{bmatrix} \quad (4.15)$$

These three matrices, complex numbers, are a representation of the order three cyclic finite group C_3. We could not write C_3 as (4.15) if Euclidean space did not exist, and we could not write C_3 as (4.14) if the complex numbers, $\mathbb{C}$, did not exist.

Group theory is now easy:
From now onward, finite group theory is easy. Yes, a finite group is a set of permutations, but, by using permutation matrix notation to represent those permutations, we can reduce finite groups to nothing more than a few sparse matrices and matrix multiplication. Actually, the 'few' gets very large very quickly; the symmetric group S_4 consists of twenty-four 4×4 permutation matrices, and it can very sensibly and most usefully be written as 24×24 permutation matrices – twenty-four of them. The concept is simple, the notation is cumbersome.

---

[12] All this about square roots of minus one is rather misleading.

Of course no one uses matrices to analyse finite groups of order 100,000 and above, but, none-the-less, group theory is about permutation matrices with matrix multiplication.

The importance of permutations:
Let us take pause to think. Is a fully fledged square matrix with entries in every position is really just a 'beefed up' set of permutation matrices? No, the algebraic matrix forms are all 'beefed up' sets of permutation matrices. General matrices do not form division algebras. The multiplication operation of every type of division algebra (type of complex numbers) is, as will be shown later, no more than matrix multiplication. But the whole of mathematics is founded upon types of numbers (division algebras), and so the whole of mathematics is founded upon matrix multiplication, and so the whole of mathematics is founded upon permutation matrices. But permutation matrices are just permutations, and matrix multiplication is just sequentially combining permutations together. It does give pause for thought.

Note: multiplying matrices of different shapes and different sizes like:

$$[a \quad b \quad c] \begin{bmatrix} d \\ e \\ f \end{bmatrix} = [ad + be + cf] \qquad (4.16)$$

is not proper multiplication. It is a duck mated with a chicked producing a wren. 'Multiplication' like (4.16) is a convenient calculative procedure; it is not multiplication.

Different representations:
We saw above that the group C_3 can be written using three 3×3 permutation matrices, (4.10), or using three 2×2 matrices, (4.14). The three 2×2 matrices, (4.14), are not permutation matrices in the sense that they do not have a single 1 in each row and in each column, but they are permutation matrices in the sense that they are in one-to-one correspondence with the 3×3 permutation matrices, (4.10). We say that

the 3×3 permutation matrices, (4.10), are a 3-dimensional representation of the finite group C_3, and we say that the 2×2 matrices, (4.14), are a 2-dimensional representation of the finite group C_3. There are other representations of the finite group C_3; for example, we have three 6×6 permutation matrices which are a 6-dimensional representation of the finite group C_3. This is:

$$C_3 =$$

$$\begin{bmatrix} 1 & 0 & 0 & 0 & 0 & 0 \\ 0 & 1 & 0 & 0 & 0 & 0 \\ 0 & 0 & 1 & 0 & 0 & 0 \\ 0 & 0 & 0 & 1 & 0 & 0 \\ 0 & 0 & 0 & 0 & 1 & 0 \\ 0 & 0 & 0 & 0 & 0 & 1 \end{bmatrix}, \begin{bmatrix} 0 & 1 & 0 & 0 & 0 & 0 \\ 0 & 0 & 1 & 0 & 0 & 0 \\ 1 & 0 & 0 & 0 & 0 & 0 \\ 0 & 0 & 0 & 0 & 1 & 0 \\ 0 & 0 & 0 & 0 & 0 & 1 \\ 0 & 0 & 0 & 1 & 0 & 0 \end{bmatrix}, \begin{bmatrix} 0 & 0 & 1 & 0 & 0 & 0 \\ 1 & 0 & 0 & 0 & 0 & 0 \\ 0 & 1 & 0 & 0 & 0 & 0 \\ 0 & 0 & 0 & 0 & 0 & 1 \\ 0 & 0 & 0 & 1 & 0 & 0 \\ 0 & 0 & 0 & 0 & 1 & 0 \end{bmatrix}$$

(4.17)

In this book, we will use the term 'permutation matrix' to refer to only matrices with a single 1 in each row and a single 1 in each column.

Inverse permutation matrices:
Above, (3.16), we saw that the inverse of a permutation is calculated by just swapping the top and bottom rows; we can rearrange the result for neatness if we like.

Given a permutation matrix, we can read off the permutation as the positions of the 1's in the matrix; for example:

$$\begin{pmatrix} 1 & 2 & 3 & 4 \\ 3 & 1 & 4 & 2 \end{pmatrix}$$

$$\begin{bmatrix} 0 & 1 & 0 & 0 \\ 0 & 0 & 0 & 1 \\ 1 & 0 & 0 & 0 \\ 0 & 0 & 1 & 0 \end{bmatrix} \qquad (4.18)$$

We then calculate the inverse of the permutation by swapping the rows:

$$\begin{pmatrix} 1 & 2 & 3 & 4 \\ 3 & 1 & 4 & 2 \end{pmatrix}^{-1} = \begin{pmatrix} 3 & 1 & 4 & 2 \\ 1 & 2 & 3 & 4 \end{pmatrix} = \begin{pmatrix} 1 & 2 & 3 & 4 \\ 2 & 4 & 1 & 3 \end{pmatrix} \quad (4.19)$$

From this, we can write the permutation matrix that is the inverse of the original given permutation matrix:

$$\begin{pmatrix} 1 & 2 & 3 & 4 \\ 2 & 4 & 1 & 3 \end{pmatrix}$$

$$\begin{bmatrix} 0 & 0 & 1 & 0 \\ 1 & 0 & 0 & 0 \\ 0 & 0 & 0 & 1 \\ 0 & 1 & 0 & 0 \end{bmatrix} \quad (4.20)$$

The reader can check that the product of these two matrices, (4.18) & (4.20) is the identity matrix. With a bit of practice, we do not need the permutations; the inverse of a permutation matrix is quite obvious to anyone who can twist their head around matrix multiplication – it is just the transpose.

Note: Transposition of a matrix is in general not an algebraic operation. It is most often meaningless. However, in the case of permutation matrices, the transpose is the inverse; we flip permutations, and we transpose permutation matrices.

Permutations which leave something unchanged:
There are two permutations of the positions of three objects which swap the positions of all three objects. These are:

$$\begin{pmatrix} 1 & 2 & 3 \\ 3 & 1 & 2 \end{pmatrix} \equiv \begin{bmatrix} 0 & 1 & 0 \\ 0 & 0 & 1 \\ 1 & 0 & 0 \end{bmatrix} \qquad \begin{pmatrix} 1 & 2 & 3 \\ 2 & 3 & 1 \end{pmatrix} \equiv \begin{bmatrix} 0 & 0 & 1 \\ 1 & 0 & 0 \\ 0 & 1 & 0 \end{bmatrix} \quad (4.21)$$

Notice there is not a 1 on the leading diagonal of the associated permutation matrices.

Permutation Matrices

The identity permutation swaps no positions of three objects. The identity permutation matrix has three 1's on the leading diagonal; we have the identity permutation matrix:

$$\begin{bmatrix} 1 & 0 & 0 \\ 0 & 1 & 0 \\ 0 & 0 & 1 \end{bmatrix} \quad (4.22)$$

The other three permutations of three objects all leave one position unchanged. The corresponding permutation matrices all have a single 1 on the leading diagonal of the associated permutation matrix in the position corresponding to the unchanged position.

$$\begin{pmatrix} 1 & 2 & 3 \\ 1 & 3 & 2 \end{pmatrix} \equiv \begin{bmatrix} 1 & 0 & 0 \\ 0 & 0 & 1 \\ 0 & 1 & 0 \end{bmatrix} \qquad \begin{pmatrix} 1 & 2 & 3 \\ 2 & 1 & 3 \end{pmatrix} \equiv \begin{bmatrix} 0 & 1 & 0 \\ 1 & 0 & 0 \\ 0 & 0 & 1 \end{bmatrix}$$

$$\begin{pmatrix} 1 & 2 & 3 \\ 3 & 2 & 1 \end{pmatrix} \equiv \begin{bmatrix} 0 & 0 & 1 \\ 0 & 1 & 0 \\ 1 & 0 & 0 \end{bmatrix} \quad (4.23)$$

A 1 on the leading diagonal of a matrix in general does nothing under matrix multiplication. We have an example of a 4×4 permutation matrix:

$$\begin{pmatrix} 1 & 2 \\ 2 & 1 \end{pmatrix} \begin{pmatrix} 3 \\ 3 \end{pmatrix} \begin{pmatrix} 4 \\ 4 \end{pmatrix} \equiv \begin{bmatrix} 0 & 1 & 0 & 0 \\ 1 & 0 & 0 & 0 \\ 0 & 0 & 1 & 0 \\ 0 & 0 & 0 & 1 \end{bmatrix} \quad (4.24)$$

Looking back to how we form a permutation matrix from a permutation, (4.5), it is obvious that an unchanged position in a permutation will correspond to a 1 on the leading diagonal of the corresponding permutation matrix.

Commutativity:
Above, we saw that permutation multiplication is non-commutative. Not surprisingly, permutation matrices can be non-commutative. However, some permutations are commutative with each other. The identity permutation commutes with all other permutations just as the matrix identity commutes with all other permutation matrices. Of course, a permutation matrix commutes with its inverse just as a permutation commutes with its inverse.

Abelian and non-abelian finite groups:
There are two types of finite groups. Some finite groups are commutative, and the other finite groups are non-commutative. A set of mathematical objects, $\{X,Y\}$ is commutative if $XY = YX$. A set of mathematical objects, $\{X,Y\}$ is non-commutative if $XY \neq YX$.

An example of a commutative finite group is the C_3 subgroup of the S_3 group. We have:

$$\begin{pmatrix} 1 & 2 & 3 \\ 1 & 2 & 3 \end{pmatrix} \equiv \begin{bmatrix} 1 & 0 & 0 \\ 0 & 1 & 0 \\ 0 & 0 & 1 \end{bmatrix} \qquad \begin{pmatrix} 1 & 2 & 3 \\ 3 & 1 & 2 \end{pmatrix} \equiv \begin{bmatrix} 0 & 1 & 0 \\ 0 & 0 & 1 \\ 1 & 0 & 0 \end{bmatrix}$$

$$\begin{pmatrix} 1 & 2 & 3 \\ 2 & 3 & 1 \end{pmatrix} \equiv \begin{bmatrix} 0 & 0 & 1 \\ 1 & 0 & 0 \\ 0 & 1 & 0 \end{bmatrix}$$

(4.25)

It is a simple fact that these matrices, (4.25), all commute, as, of course, do the permutations corresponding to these matrices. An example of a non-commutative finite group is the symmetric group S_3 above, (4.3) & (4.4). The S_3 permutation matrices do not commute.

Permutation Matrices

We call the commutative groups abelian groups after the mathematician Niels Henrik Abel (1802-1829), and we call the non-commutative groups non-abelian groups[13].

Note that, although all cyclic groups like C_3 are abelian, not every abelian group is a cyclic group. There are abelian groups which are not cyclic groups.

Closure of permutation matrices:
We multiply matrices together by taking the product of a row, Row, and a column, Col; this gives the $a_{Row,Col}$ element in the product matrix. Since there is only a single 1 in each row of a permutation matrix and a single 1 in each column of a permutation matrix, any product of two permutation matrices must be another permutation matrix. The permutation matrices are a set of multiplicatively closed matrices.

There are subsets of permutation matrices which are multiplicatively closed within their own right.

Summary:
Permutations are correlated with permutation matrices as shown above in (4.5).

The order of an element of a group is the power to which that element must be raised to be equal to the identity. Every element of a group will become the identity if raised to the appropriate power.

Permutation multiplication is really just matrix multiplication. Hm! perhaps matrix multiplication is really just permutation multiplication.

There are many sets of permutation matrices which represent a given finite group. These sets can be, but need not be, of different sizes. We call these different sets of permutation matrices which represent a given finite group different representations of the given finite group.

---

[13] Presumably after the mathematician Niels Henrik Non-Abel (1802-1829).

The inverse of a permutation is found by swapping the rows of the permutation. The inverse of a permutation matrix is the transpose of that permutation matrix.

Finite groups can be commutative (abelian) or non-commutative (non-abelian).

Addendum – special relativity from permutations:
The exponential of a matrix, M, is given by:

$$\exp(M) = 1 + M + \frac{M^2}{2!} + \frac{M^3}{3!} + \ldots \quad (4.26)$$

$$\exp\left(\begin{bmatrix} 0 & 1 \\ 1 & 0 \end{bmatrix}\right) = \begin{bmatrix} 1 & 0 \\ 0 & 1 \end{bmatrix} + \begin{bmatrix} 0 & 1 \\ 1 & 0 \end{bmatrix} + \frac{\begin{bmatrix} 1 & 0 \\ 0 & 1 \end{bmatrix}}{2!} + \frac{\begin{bmatrix} 0 & 1 \\ 1 & 0 \end{bmatrix}}{3!} + \ldots$$

$$= \begin{bmatrix} \cosh 1 & \sinh 1 \\ \sinh 1 & \cosh 1 \end{bmatrix} \quad (4.27)$$

In terms of permutations, the exponential of this permutation is:

$$\exp\left(\begin{pmatrix} 1 & 2 \\ 2 & 1 \end{pmatrix}\right) = \begin{pmatrix} 1 & 2 \\ 1 & 2 \end{pmatrix} + \begin{pmatrix} 1 & 2 \\ 2 & 1 \end{pmatrix} + \frac{1}{2!}\begin{pmatrix} 1 & 2 \\ 1 & 2 \end{pmatrix} + \frac{1}{3!}\begin{pmatrix} 1 & 2 \\ 2 & 1 \end{pmatrix} + \ldots$$

$$= \begin{bmatrix} \cosh 1 & \sinh 1 \\ \sinh 1 & \cosh 1 \end{bmatrix}$$

(4.28)

If we multiply the permutation by a real number, χ, we get:

$$\exp\left(\chi \begin{pmatrix} 1 & 2 \\ 2 & 1 \end{pmatrix}\right) = \begin{bmatrix} \cosh \chi & \sinh \chi \\ \sinh \chi & \cosh \chi \end{bmatrix} \quad (4.29)$$

This, (4.29), is the rotation matrix of the 2-dimensional space-time of special relativity. Rotation in 2-dimensional space-time is just change of velocity. We have:

Permutation Matrices

$$\begin{bmatrix} \cosh \chi & \sinh \chi \\ \sinh \chi & \cosh \chi \end{bmatrix} = \begin{bmatrix} \gamma & \gamma v \\ \gamma v & \gamma \end{bmatrix}$$

$$\gamma = \frac{1}{\sqrt{1 - \frac{v^2}{c^2}}}$$ (4.30)

We see that the whole concept of velocity, that is movement through space and time, is no more that a combination of a real number and a permutation. Quite profound, methinks.

Chapter 5

The Symmetric Groups

Our first type of finite group is the symmetric groups. These are groups denoted by S_n. The symmetric group, S_n, is the set of every $n \times n$ permutation matrix. An alternative but equivalent definition is to say that the symmetric group, S_n, is the set of every permutation of n objects. For example, the symmetric group S_4 is the set of the twenty-four order-4 permutations; these are the twenty-four 4×4 permutation matrices (or twenty-four permutations of four objects, if you prefer).

Clearly, there are finite groups other than the symmetric groups; we have already found a cyclic subgroup, C_3, in S_3.

Subgroups of symmetric groups:
All finite groups contain within themselves subgroups which are finite groups; these are closed sets of permutations in their own right. One such subgroup is the identity which is a finite group of only one element. Another such subgroup is the whole group itself. As well as these two subgroups, most, but not all, finite groups contain subgroups of lesser order than themselves but of order greater than one. Such subgroups are called proper subgroups.

The symmetric groups in general have proper subgroups; an example is the symmetric group S_3 which has an order three cyclic group, C_3, as a subgroup. The order five cyclic group C_5 is not a proper subgroup of the order six symmetric group S_3, but it is a proper subgroup of the order 120 symmetric group S_5. In general, the finite group S_n also has other proper subgroups, for example, every symmetric group of order n has an alternating group of order $\frac{n}{2}$ as a proper subgroup.

Every type of finite group is a subgroup of some symmetric group. Although we have not yet shown it, every finite group of order n is a set of $n \times n$ permutation matrices; as such, it is some subset of the set of all $n \times n$ permutation matrices. Thus every finite group of order n is a subgroup of the symmetric group S_n. As we have said above, S_n is of order n-factorial, $n!$.

Order of the symmetric groups:
How many permutation matrices are there of a given size? Let us go through the procedure of constructing a permutation matrix. Well, in a $n \times n$ matrix, there are n positions in the top row where we could place a 1; that is n choices. We choose one of these positions. Similarly, there are n positions in the second row where we could place a 1, but we cannot use the position which is in the same column as the 1 in the top row. Remember, a permutation matrix has a single 1 in each row and a single 1 in each column. This means there are $(n-1)$ positions in the second row where we could place a 1; that is $(n-1)$ choices. We choose one of these $(n-1)$ positions. Similarly, there are $(n-2)$ positions in the third row where we could place a 1; we have to avoid the two columns which already have a 1 in them. We choose one of these $(n-2)$ positions. Eventually, we get to the bottom row; there is only one position available to us. We have now constructed a single permutation matrix. How many different ways can we do this? The answer is to multiply together the number of choices on each row of the matrix:

$$n \times (n-1) \times (n-2) \times \ldots \times 3 \times 2 \times 1 = n! \quad (5.1)$$

Since the symmetric group, S_n contains all $n \times n$ permutation matrices, the order of S_n is $n!$.

We now have an infinite set of finite groups; these are the symmetric groups. In each case, we know the order of the symmetric group. We know that these symmetric groups have proper subgroups and we know the nature of some of the subgroups, but we do not know every proper subgroup that is within each symmetric group – we do not know the entire subgroup structure of the symmetric groups.

Symmetric groups contain lower order symmetric groups:
The symmetric group S_n has as proper subgroups all the symmetric groups of lesser order than itself.

Consider the S_3 permutations:

$$\begin{pmatrix} 1 & 2 & 3 \\ 3 & 1 & 2 \end{pmatrix} \qquad \begin{pmatrix} 1 & 2 & 3 \\ 2 & 3 & 1 \end{pmatrix} \qquad (5.2)$$

Now consider the S_4 permutations:

$$\begin{pmatrix} 1 & 2 & 3 & 4 \\ 3 & 1 & 2 & 4 \end{pmatrix} \qquad \begin{pmatrix} 1 & 2 & 3 & 4 \\ 2 & 3 & 1 & 4 \end{pmatrix} \qquad (5.3)$$

These pairs of permutations, (5.2) & (5.3), are the effectively same pair of permutations. Similarly, the other four permutations within S_3 are also contained within S_4. Similar reasoning leads to the whole of the finite group S_{n-1} being contained within the finite group S_n. But each S_{n-1} is a finite group in its own right. Thus all symmetric groups contain all lesser order symmetric groups as proper subgroups. This implies that all the subgroups of the lesser order symmetric groups are also subgroups of the higher order symmetric groups; for example, since S_3 has a C_3 subgroup, every $S_n : n > 3$ has at least one C_3 subgroup.

Put less mathematically, this inclusion of lesser order symmetric groups within the higher order symmetric group is just permutations of balls in which the position of one of the balls remains unchanged. If we have four coloured balls and we permute only three of them, we get a closed set of six permutations which are six of the twenty-four permutations of four coloured balls.

The Symmetric Groups

Summary:

We have found our first infinite set of a type of finite groups – the symmetric groups.

A symmetric group is a complete set of all possible permutations of n objects. This corresponds to a complete set of all $n \times n$ permutation matrices.

We know the order of each symmetric group, but we know only part of the subgroup structure of the symmetric groups.

Finite Groups – A Simple Introduction

Chapter 6

Other Representations and Cayley Tables

Above, (4.3) & (4.4), we presented the order six finite group S_3 as six 3×3 permutation matrices. We could have presented the same finite group using six 6×6 permutation matrices. We have:

$$\begin{bmatrix} 1 & 0 & 0 & 0 & 0 & 0 \\ 0 & 1 & 0 & 0 & 0 & 0 \\ 0 & 0 & 1 & 0 & 0 & 0 \\ 0 & 0 & 0 & 1 & 0 & 0 \\ 0 & 0 & 0 & 0 & 1 & 0 \\ 0 & 0 & 0 & 0 & 0 & 1 \end{bmatrix}, \begin{bmatrix} 0 & 1 & 0 & 0 & 0 & 0 \\ 0 & 0 & 1 & 0 & 0 & 0 \\ 1 & 0 & 0 & 0 & 0 & 0 \\ 0 & 0 & 0 & 0 & 1 & 0 \\ 0 & 0 & 0 & 0 & 0 & 1 \\ 0 & 0 & 0 & 1 & 0 & 0 \end{bmatrix}, \begin{bmatrix} 0 & 0 & 1 & 0 & 0 & 0 \\ 1 & 0 & 0 & 0 & 0 & 0 \\ 0 & 1 & 0 & 0 & 0 & 0 \\ 0 & 0 & 0 & 0 & 0 & 1 \\ 0 & 0 & 0 & 1 & 0 & 0 \\ 0 & 0 & 0 & 0 & 1 & 0 \end{bmatrix}$$

(6.1)

$$\begin{bmatrix} 0 & 0 & 0 & 1 & 0 & 0 \\ 0 & 0 & 0 & 0 & 0 & 1 \\ 0 & 0 & 0 & 0 & 1 & 0 \\ 1 & 0 & 0 & 0 & 0 & 0 \\ 0 & 0 & 1 & 0 & 0 & 0 \\ 0 & 1 & 0 & 0 & 0 & 0 \end{bmatrix}, \begin{bmatrix} 0 & 0 & 0 & 0 & 1 & 0 \\ 0 & 0 & 0 & 1 & 0 & 0 \\ 0 & 0 & 0 & 0 & 0 & 1 \\ 0 & 1 & 0 & 0 & 0 & 0 \\ 1 & 0 & 0 & 0 & 0 & 0 \\ 0 & 0 & 1 & 0 & 0 & 0 \end{bmatrix}, \begin{bmatrix} 0 & 0 & 0 & 0 & 0 & 1 \\ 0 & 0 & 0 & 0 & 1 & 0 \\ 0 & 0 & 0 & 1 & 0 & 0 \\ 0 & 0 & 1 & 0 & 0 & 0 \\ 0 & 1 & 0 & 0 & 0 & 0 \\ 1 & 0 & 0 & 0 & 0 & 0 \end{bmatrix}$$

(6.2)

This, (6.1) & (6.2) is the finite symmetric group S_3.

There is obviously a correlation between the above six S_3 6×6 matrices (6.1) & (6.2) and the six 3×3 S_3 matrices (4.3) & (4.4). For a start, the two identity matrices are correlated together. If we use different variables in the three 3×3 matrices, (4.3) & (4.4), of the group C_3, we have:

Other Representations and Cayley Tables

$$\begin{bmatrix} a & 0 & 0 \\ 0 & a & 0 \\ 0 & 0 & a \end{bmatrix} + \begin{bmatrix} 0 & b & 0 \\ 0 & 0 & b \\ b & 0 & 0 \end{bmatrix} + \begin{bmatrix} 0 & 0 & c \\ c & 0 & 0 \\ 0 & c & 0 \end{bmatrix} = \begin{bmatrix} a & b & c \\ c & a & b \\ b & c & a \end{bmatrix} \quad (6.3)$$

Doing the same with the three 6×6 matrices, (6.1) gives:

$$\begin{bmatrix} a & b & c & 0 & 0 & 0 \\ c & a & b & 0 & 0 & 0 \\ b & c & a & 0 & 0 & 0 \\ 0 & 0 & 0 & a & b & c \\ 0 & 0 & 0 & c & a & b \\ 0 & 0 & 0 & b & c & a \end{bmatrix} \quad (6.4)$$

We see that we have just copied two copies of the 3×3 matrix on to the leading diagonal of the 6×6 matrix. This is a 6-dimensional representation of the order three cyclic group C_3.

Doing the same with the other three 3×3 matrices, we get:

$$\begin{bmatrix} d & 0 & 0 \\ 0 & 0 & d \\ 0 & d & 0 \end{bmatrix} + \begin{bmatrix} 0 & e & 0 \\ e & 0 & 0 \\ 0 & 0 & e \end{bmatrix} + \begin{bmatrix} 0 & 0 & f \\ 0 & f & 0 \\ f & 0 & 0 \end{bmatrix} = \begin{bmatrix} d & e & f \\ e & f & d \\ f & d & e \end{bmatrix} \quad (6.5)$$

We note that these three matrices, (6.5), each have an element on the leading diagonal. The leading diagonal is reserved for the identity matrix. We fit the 3×3 matrix, (6.5), into the off-diagonal position of (6.4):

$$S_3 = \begin{bmatrix} a & b & c & d & e & f \\ c & a & b & e & f & d \\ b & c & a & f & d & e \\ d & e & f & a & b & c \\ e & f & d & c & a & b \\ f & d & e & b & c & a \end{bmatrix} \quad (6.6)$$

47

Separating out the individual variables will give the six 6×6 matrices above, (6.1) & (6.2).

There are obviously a lot more 6×6 permutation matrices than just these six, (6.1) & (6.2); there are $6! = 720$ of them. The entire set of 6×6 permutation matrices is the symmetric finite group S_6. The above six 6×6 permutation matrices which are the S_3 finite group are a proper subgroup of S_6.

Above, (4.4), we found that only three of the 3×3 permutation matrices form the order three finite cyclic group C_3.

It is a fact that any finite group of order n can be written as n $n \times n$ permutation matrices. Many groups can be written as a set of permutation matrices of lesser size than the order of the group; an example of this is the representation of the order six finite group S_3 as six 3×3 matrices.

The size of the permutation matrices representing a finite group is not limited to being equal to the order of the group or less. The order two finite group C_2, which is the two 2×2 permutation matrices above, (4.2), can be represented as the two 4×4 permutation matrices:

$$\begin{bmatrix} 1 & 0 & 0 & 0 \\ 0 & 1 & 0 & 0 \\ 0 & 0 & 1 & 0 \\ 0 & 0 & 0 & 1 \end{bmatrix} \& \begin{bmatrix} 0 & 1 & 0 & 0 \\ 1 & 0 & 0 & 0 \\ 0 & 0 & 0 & 1 \\ 0 & 0 & 1 & 0 \end{bmatrix} \quad (6.7)$$

There are other such representations of C_2 using 4×4 permutation matrices. Of course, every representation will include the identity matrix of the appropriate size because every group has an identity element.

The adjoint representation:

We refer to a set of $n \times n$ permutation matrices representing a group of order n as the adjoint representation of that finite group[14]. Obviously, there will be n of these $n \times n$ permutation matrices in the adjoint representation of a finite group. The adjoint representation is also called the regular representation.

What is remarkable is that the n permutation matrices in the adjoint representation fit together perfectly to form a matrix in which every element is a one. For example, add the six 6×6 matrices representing S_3, (6.1) & (6.2), and we get the matrix:

$$\begin{bmatrix} 1 & 1 & 1 & 1 & 1 & 1 \\ 1 & 1 & 1 & 1 & 1 & 1 \\ 1 & 1 & 1 & 1 & 1 & 1 \\ 1 & 1 & 1 & 1 & 1 & 1 \\ 1 & 1 & 1 & 1 & 1 & 1 \\ 1 & 1 & 1 & 1 & 1 & 1 \end{bmatrix} \qquad (6.8)$$

We can do the same with the three 3×3 matrices representing the finite group C_3, (4.25), to get:

$$\begin{bmatrix} 1 & 1 & 1 \\ 1 & 1 & 1 \\ 1 & 1 & 1 \end{bmatrix} \qquad (6.9)$$

This phenomenon is general.

Since the inverse, transpose permutation matrix, of every element of a group is also an element of a group, then the separate elements of the group are extracted from the adjoint representation in 'pairs' which are the element and its inverse; however, sometimes, there is only one element in the 'pair'. Each element of the 'pair' is the transpose of the other element, but some permutation matrices are their own transposes

---

[14] This is a terminology taken from Lie group theory. It is not commonly seen within finite group theory, but it fits very well.

– they are symmetric matrices. These 'only one in a pair' matrices are their own inverses and so, other than the identity, these elements are of order two. We have an example in which we have previously extracted the identity:

$$\begin{bmatrix} 0 & 1 & 1 & 1 & 1 & 1 \\ 1 & 0 & 1 & 1 & 1 & 1 \\ 1 & 1 & 0 & 1 & 1 & 1 \\ 1 & 1 & 1 & 0 & 1 & 1 \\ 1 & 1 & 1 & 1 & 0 & 1 \\ 1 & 1 & 1 & 1 & 1 & 0 \end{bmatrix} \rightarrow \begin{bmatrix} 0 & 1 & 1 & 1 & 0 & 1 \\ 1 & 0 & 1 & 0 & 1 & 1 \\ 1 & 1 & 0 & 1 & 1 & 0 \\ 1 & 0 & 1 & 0 & 1 & 1 \\ 0 & 1 & 1 & 1 & 0 & 1 \\ 1 & 1 & 0 & 1 & 1 & 0 \end{bmatrix} + \begin{bmatrix} 0 & 0 & 0 & 0 & 1 & 0 \\ 0 & 0 & 0 & 1 & 0 & 0 \\ 0 & 0 & 0 & 0 & 0 & 1 \\ 0 & 1 & 0 & 0 & 0 & 0 \\ 1 & 0 & 0 & 0 & 0 & 0 \\ 0 & 1 & 0 & 0 & 0 & 0 \end{bmatrix}$$
(6.10)

Symmetric matrices have real eigenvalues and orthogonal eigenvectors just like the hermitian matrices of quantum mechanics. In fact, hermitian matrices are just symmetric matrices in obscure notation.

The manner in which we extract individual permutation matrices dictates which group we get.

The adjoint representation is not unique for groups of order greater than three; for example, there are three adjoint representations of the order four cyclic group C_4. We identify the separate permutation matrices by labelling each of them with a different lower case letter. We then add the differently labelled permutation matrices to display the adjoint representation of the finite group in a compact format. In the case of the finite group C_4, the three different adjoint representations are:

$$C_4^A \equiv \begin{bmatrix} a & b & c & d \\ d & a & b & c \\ c & d & a & b \\ b & c & d & a \end{bmatrix} \quad C_4^B \equiv \begin{bmatrix} a & b & c & d \\ b & a & d & c \\ d & c & a & b \\ c & d & b & a \end{bmatrix} \quad C_4^C \equiv \begin{bmatrix} a & b & c & d \\ c & a & d & b \\ b & d & a & c \\ d & c & b & a \end{bmatrix}$$
(6.11)

By looking for the transposes in each of (6.11), we can see that in C_4^A the element denoted by c is its own inverse and therefore an element of order two but that in C_4^B the element denoted by c is the inverse of the element denoted by d.

The group S_4 has three C_4 subgroups corresponding to the above, (6.11). As sets of permutations, by simply reading the permutations from each of the three adjoint representations of C_4 above, (6.11), the three permutation forms of C_4 are:

$$C_4^A = \left\{ \begin{pmatrix} 1 & 2 & 3 & 4 \\ 1 & 2 & 3 & 4 \end{pmatrix}, \begin{pmatrix} 1 & 2 & 3 & 4 \\ 4 & 1 & 2 & 3 \end{pmatrix}, \begin{pmatrix} 1 & 2 & 3 & 4 \\ 3 & 4 & 1 & 2 \end{pmatrix}, \begin{pmatrix} 1 & 2 & 3 & 4 \\ 2 & 3 & 4 & 1 \end{pmatrix} \right\}$$

(6.12)

$$C_4^B = \left\{ \begin{pmatrix} 1 & 2 & 3 & 4 \\ 1 & 2 & 3 & 4 \end{pmatrix}, \begin{pmatrix} 1 & 2 & 3 & 4 \\ 2 & 1 & 4 & 3 \end{pmatrix}, \begin{pmatrix} 1 & 2 & 3 & 4 \\ 4 & 3 & 1 & 2 \end{pmatrix}, \begin{pmatrix} 1 & 2 & 3 & 4 \\ 3 & 4 & 2 & 1 \end{pmatrix} \right\}$$

(6.13)

$$C_4^C = \left\{ \begin{pmatrix} 1 & 2 & 3 & 4 \\ 1 & 2 & 3 & 4 \end{pmatrix}, \begin{pmatrix} 1 & 2 & 3 & 4 \\ 3 & 1 & 4 & 2 \end{pmatrix}, \begin{pmatrix} 1 & 2 & 3 & 4 \\ 2 & 4 & 1 & 3 \end{pmatrix}, \begin{pmatrix} 1 & 2 & 3 & 4 \\ 4 & 3 & 2 & 1 \end{pmatrix} \right\}$$

(6.14)

The group S_4, which is the set of all order four permutations, has three C_4 subgroups precisely because there are three different sets of order four permutations which form the group C_4. We might say that there are three copies of C_4 within the set of order 4 permutations.

The order four finite cyclic group C_4 has a single proper C_2 subgroup. If the reader squares each of the permutations from one of the above sets, (6.12) to (6.14), the reader will find that, in each set, one of the three non-identity permutations squares to the identity (is of order two); this is the single order two element of the C_2 subgroup of C_4. Looking at (6.11) above, we see that the three representations of C_4 above, (6.11), differ from each other in which particular variable, $\{b$ or c or $d\}$, is the generator of the C_2 subgroup within C_4.

The Cayley table:

If we label each of the permutation matrices in the adjoint representation of a finite group by a different lower case real letter and we add the matrices, we get the multiplication table of the finite group. For example, the multiplication table of C_3 is:

$$C_3 \sim \begin{bmatrix} a & b & c \\ c & a & b \\ b & c & a \end{bmatrix} \qquad (6.15)$$

This says that, if we choose an element on the top row in the xth column and another element in the leftmost column in the yth row, then the product of those two elements of the group in the order *top row × leftmost column* is the element in the main body of the table at the position, (x, y) that correlates to the positions of the chosen elements in the top row and leftmost column. For examples, looking at (6.15), we have $b \times c = a$ and $b \times b = c$ and $c \times c = b$.

The multiplication table of a group is called the Cayley table of the group after the mathematician Arthur Cayley[15].

There are different ways to write the Cayley table of a finite group by ordering the elements in the top row or the leftmost column differently. We have chosen to write the Cayley table with the identity elements on the leading diagonal. This is called the Standard Form Cayley table. It is the standard form of the Cayley table which always emerges from adding the labelled permutation matrices of the adjoint representation of the finite group. The order four cyclic group, C_4, has three Standard Form Cayley tables – see (6.11).

It is easy to form different Cayley tables. Provided that we keep the identity in the top left hand corner, we just swap a row or swap a column or both. For example, swapping a row and a column of (6.15) will give:

---

[15] Arthur Cayley : Phil. Mag. Vii(4) 1854

Other Representations and Cayley Tables

$$C_3 \sim \begin{bmatrix} a & c & b \\ c & b & a \\ b & a & c \end{bmatrix} \sim \begin{bmatrix} a & c & b \\ b & a & c \\ c & b & a \end{bmatrix} \qquad (6.16)$$

Notice how the left-most of these, (6.16), is a symmetric matrix; that is reflectively symmetric across the leading diagonal[16]. It is a fact that all commutative groups have a Cayley table which can be put into a symmetrical form by swapping rows and columns holding the identity in the top left-hand corner.

Why can we write the Cayley table of an abelian (commutative) group as a symmetrical table? Because $b \times c = c \times b$, and so $(x, y) = (y, x)$ within the Cayley table. The Standard form Cayley table is often not symmetrical, see (6.15), even though a group is abelian. This is one of the reasons why the Standard Form Cayley table is not the only form of Cayley table used in the literature.

Take care with Cayley tables:
The Cayley table gives the product of two elements of the group in the order *top row × left column*. For a commutative group like C_3 above, (6.16), we have *top row × left column = left column × top row*, but for a non-commutative group, this is not so. It is a convention that the Cayley table presents the products of two elements of a group in the *top row × left column* order.

Not everything is a Cayley table:
One of the defining attributes of a finite group is that the multiplication of the elements of the group must be associative[17]. Combining permutations is always associative, as is matrix multiplication.

---

[16] The leading diagonal runs from the top left-hand corner to the bottom right-hand corner.
[17] Not all 'multiplication' is associative. Octonians are non-associative.

Defining a finite group as a set of permutations automatically leads to this associativity of finite groups.

Consider:

$$\begin{bmatrix} a & b & c & d & e \\ b & a & e & c & d \\ c & d & a & e & b \\ d & e & b & a & c \\ e & c & d & b & a \end{bmatrix} \quad (6.17)$$

This, (6.17), cannot be a multiplication table of a finite group because, using the table, (6.17), we have:

$$\begin{aligned} (bc)d &= ed = b \\ b(cd) &= be = d \end{aligned} \quad (6.18)$$

The multiplication is not associative.

Summary:

A finite group is a closed set of permutations. We can take the view that a finite group is a set of permutation matrices.

We have met the adjoint representation of a finite group of order n as the sum of a set of $n \times n$ permutation matrices. We have seen that, for finite groups of order more than three, there is more than one adjoint representation.

The adjoint representation is very important because it leads directly to all the various forms of division algebras. For this reason, the labelled and summed adjoint representation is often called the algebraic matrix form of the finite group.

We have met the Standard Form Cayley table of a finite group of order n as the sum of a set of labelled $n \times n$ permutation matrices. We have met the Cayley table of a group in both its Standard Form and other forms.

The adjoint representation is pretty much the same thing as the Standard Form Cayley table.

Chapter 7

An Algebraic View

The order of an element:
The nature of every element of a finite group, that is every permutation, is such that when it is raised to the appropriate power it will equal the identity. The appropriate power is called the order of that element.

Roots of plus unity:
The identity of a set of permutations is basically equivalent to the number one[18]. The identity of a set of permutation matrices is the number one, at least up to algebraic isomorphism. We can therefore view the non-identity elements of the group as roots of plus one. We might describe the order six symmetric finite group, S_3, as:

$$S_3 \equiv 1 + 3\sqrt[2]{+1} + 2\sqrt[3]{+1} \qquad (7.1)$$

And the order three cyclic group, C_3, as:

$$C_3 \equiv 1 + 2\sqrt[3]{+1} \qquad (7.2)$$

A similar view can be taken of every finite group[19]. We present a few examples:

$$C_6 \equiv 1 + \sqrt[2]{+1} + 2\sqrt[3]{+1} + 2\sqrt[6]{+1}$$
$$C_7 \equiv 1 + 6\sqrt[7]{+1} \qquad (7.3)$$
$$Q_8 \equiv 1 + \sqrt[2]{+1} + 6\sqrt[4]{+1}$$

---

[18] Technically, the identity permutation is, albeit indirectly, algebraically isomorphic to the number one.
[19] See : Dennis Morris : Complex Numbers The Higher Dimensional Forms.

An Algebraic View

Division Algebras from the finite groups:

We begin with the Standard Form Cayley Table of a finite group as formed by adding the labelled permutation matrices of the adjoint representation of the given finite group. We convert the Standard Form Cayley table into the algebraic matrix form of the group by seeing each of the lower case labels as a real variable.

We take the matrix exponential of the algebraic matrix form, Standard Form Cayley Table, and we get a type of complex numbers. For example, using the group C_3:

$$\exp\left(\begin{bmatrix} a & b & c \\ c & a & b \\ b & c & a \end{bmatrix}\right) = \begin{bmatrix} e^a & 0 & 0 \\ 0 & e^a & 0 \\ 0 & 0 & e^a \end{bmatrix} \begin{bmatrix} v_A(b,c) & v_B(b,c) & v_C(b,c) \\ v_C(b,c) & v_A(b,c) & v_B(b,c) \\ v_B(b,c) & v_C(b,c) & v_A(b,c) \end{bmatrix}$$

(7.4)

This is a type of 3-dimensional complex numbers in polar form. The e^a matrix is the radial part of the polar form and the $v_i(b,c)$ matrix is the 3-dimensional rotation matrix which is the angular part of the polar form of these 3-dimensional complex numbers[20]. The individual $v_i(b,c)$ are the 3-dimensional trigonometric functions associated with this type of 3-dimensional complex number.

Every finite group holds within it at least one[21] type of higher dimensional complex numbers and most often many types of higher dimensional complex numbers. The other different types of complex numbers are derived by scattering[22] minus signs throughout the algebraic matrix form and requiring multiplicative closure of form. An example is the commonly known two types of 2-dimensional complex numbers derived from the order two cyclic finite group C_2.

---

[20] For more details, see : Dennis Morris : Complex Numbers The Higher Dimensional Forms

[21] The order eight dicyclic group holds 128 copies of a single type of 8-dimensional complex numbers.

[22] It is a little more formal than mere 'scattering'.

We have:

$$C_2 = \left\{ \begin{bmatrix} 1 & 0 \\ 0 & 1 \end{bmatrix}, \begin{bmatrix} 0 & 1 \\ 1 & 0 \end{bmatrix} \right\} \to \begin{bmatrix} a & b \\ b & a \end{bmatrix} \quad (7.5)$$

This leads to:

$$\exp\left(\begin{bmatrix} a & b \\ b & a \end{bmatrix}\right) = \begin{bmatrix} e^a & 0 \\ 0 & e^a \end{bmatrix} \begin{bmatrix} \cosh b & \sinh b \\ \sinh b & \cosh b \end{bmatrix} = \mathbb{S} \quad (7.6)$$

This is the hyperbolic complex numbers which are 2-dimensional space-time.

We also have:

$$\exp\left(\begin{bmatrix} a & b \\ -b & a \end{bmatrix}\right) = \begin{bmatrix} e^a & 0 \\ 0 & e^a \end{bmatrix} \begin{bmatrix} \cos b & \sin b \\ -\sin b & \cos b \end{bmatrix} = \mathbb{C} \quad (7.7)$$

This is the Euclidean complex numbers which are 2-dimensional Euclidean space. These are commonly simply called the complex numbers, $\mathbb{C}$.

Whether it is better to think of finite groups as being closed sets of permutations or as closed sets of algebraic roots of plus unity is entirely a matter of opinion. We see how closely numbers are tied to permutations.

A word of warning:
Any set of roots of plus one which includes a square root of plus one, as many Clifford algebras do, cannot be a division algebra because it contains zero divisors. We have:

$$(1+\sqrt{+1})(1-\sqrt{+1}) = 1 - 1 = 0 \quad (7.8)$$

We see that the product of two non-zero elements of the algebra is zero.

The square roots of plus one are easily dealt with by taking the exponential of the algebraic matrix form of the prospective algebra as we did above, (7.4). This solves the zero divisors problem; it also solves related problems such as absence of multiplicative inverse.

Clifford algebras reformulated:

Similarly, we can take the exponential of any Clifford algebra and thereby render it as a division algebra by ridding it of its zero divisors.[23]

The Clifford algebras were discovered, in their non-exponential forms, by William Kingdon Clifford (1845-1879). They have a role in particle physics.

---

[23] See : Dennis Morris : The Naked Spinor

Chapter 8

The Group Axioms

Most books on finite group theory start with a list of the four group axioms. These often appear on the first page of the book. Clearly, we are a little slow in our presentation of the group axioms.

There are four group axioms. These axioms are:

a) There is an identity element: if b is any element of the group, then there is another element of the group, I, such that $bI = Ib = b$ and the element I is the one and same element for every element of the group.
b) There is closure: if $\{b,c\}$ are any two elements of the group, then so are the products bc & cb.
c) Every element has an inverse within the group: if b is an element of the group, then the group also has an element b^{-1} such that $bb^{-1} = b^1 b = I$.
d) The multiplication operation is associative: $(ab)c = a(bc)$. This means that we multiply the two elements inside the brackets before we involve the other element.

We know what an identity element is, and we already know that every finite group has an identity element. It is usual, though trivial, to point out that there is only one identity element in any finite group.

Suppose there were two identities, $\{e,e'\}$. Then, because these are identities, we have:

$$ee' = e' \quad \& \quad ee' = e \quad \text{so} \quad e = e' \qquad (8.1)$$

We know what closure is, and we already know that a finite group is a closed set of permutations, or a closed set of permutation matrices, if you prefer.

The Group Axioms

The inverse element:

Within the finite group C_3, we have the product:

$$\begin{bmatrix} 0 & 1 & 0 \\ 0 & 0 & 1 \\ 1 & 0 & 0 \end{bmatrix} \begin{bmatrix} 0 & 0 & 1 \\ 1 & 0 & 0 \\ 0 & 1 & 0 \end{bmatrix} = \begin{bmatrix} 1 & 0 & 0 \\ 0 & 1 & 0 \\ 0 & 0 & 1 \end{bmatrix} \quad (8.2)$$

$$bc = a$$

In the permutations above, (3.1), CAB is the inverse of BCA.

We see that the $b \& c$ permutation matrices in (8.2) are the inverses of each other:

$$\begin{bmatrix} 0 & 1 & 0 \\ 0 & 0 & 1 \\ 1 & 0 & 0 \end{bmatrix}^{-1} = \begin{bmatrix} 0 & 0 & 1 \\ 1 & 0 & 0 \\ 0 & 1 & 0 \end{bmatrix} \quad (8.3)$$

Some elements of a finite group are their own inverses (like minus one); for example, in the finite group C_2, (4.2), we have:

$$\begin{bmatrix} 0 & 1 \\ 1 & 0 \end{bmatrix} \begin{bmatrix} 0 & 1 \\ 1 & 0 \end{bmatrix} = \begin{bmatrix} 1 & 0 \\ 0 & 1 \end{bmatrix} \quad (8.4)$$

It is simply a fact that if you begin with the permutation ABC and change that to some other permutation, then you can always change the new permutation back to ABC. Think of permuting three coloured balls in a line.

Every element in every closed set of permutations has such an inverse, and so we declare the necessity for an inverse of each element to be a group axiom. The inverse of the identity is the identity.

The inverse of each element in a group is unique. Suppose both $\{y, z\}$ were both inverses of the element x. Then we have:

$$
\begin{aligned}
y &= ey && e \text{ is the identity of the group} \\
&= (zx)y && \text{since } z \text{ is the inverse of } x \\
&= z(xy) && \text{multiplication is associative} && (8.5) \\
&= ze && \text{since } y \text{ is an inverse of } x \\
&= z && \text{since } e \text{ is the identity}
\end{aligned}
$$

Associativity:
Non-associativity gets an unfair press. Associativity insists that the order in which we evaluate a product of three elements of a group is irrelevant. In the associative product abc, if, keeping the order the same throughout, we first multiply a by b and then multiply the result of this calculation by c, then we get the same answer as if we first multiplied b by c and then multiplied the result of this calculation by a. In symbols: $(ab)c = a(bc)$.

Mathematicians accept non-commutative elements within a finite group such that $bc \neq cb$. This means that there are two products of a pair of elements. Non-associativity means the same; we just get two products of three elements, and yet non-associativity is frowned upon by mathematicians. It seems very unfair.

Well, above we have given the traditional four group axioms. If you want to prove that a set of objects is a group, you simply show that the set of objects satisfies the four group axioms listed above.

Too many types of groups:
One of the confusions of group theory is that there are many different types of mathematical objects which satisfy all the group axioms, and so these objects are all groups, not finite groups, but still groups.

Rotation groups:
An example is a rotation matrix:

The Group Axioms

$$\begin{bmatrix} \cos\theta & \sin\theta \\ -\sin\theta & \cos\theta \end{bmatrix} \tag{8.6}$$

There is a different rotation matrix for every value of θ. There are an infinite number of such rotation matrices. This infinite set of rotation matrices, as with all rotation matrices of all dimensions, satisfies all four group axioms.

1) Identity: The identity is the rotation matrix with $\theta = 0$.
2) Closure: Two successive rotations make a rotation.
3) Inverse: The inverse of the θ rotation matrix is the rotation matrix with $-\theta$.
4) Associativity: Matrices are associative.

Thus, the rotation matrix, (8.6), is a group, but it is not a finite group. We say that it is a continuous group of infinite order. A rotation matrix is not a set of permutations, therefore it is not a finite group.

Actually, all rotation matrices are formed by taking the exponential of the sum of all permutation matrices within a finite group except the identity. Above, (7.4), is an example of this. Another example is:

$$\exp\left(\begin{bmatrix} 0 & b \\ b & 0 \end{bmatrix}\right)\begin{bmatrix} \cosh b & \sinh b \\ \sinh b & \cosh b \end{bmatrix} \tag{8.7}$$

Homotopy groups and braid groups:
Then there are homotopy groups which are associated with the number of holes in a topological space - nothing to do with finite groups. Then there are braid groups which are associated with knot theory – a tenuous connection to finite groups. Then there are many more types of objects also called groups because they satisfy the four group axioms.

There are so many disparate objects bearing the name group that the word 'group' is too much overworked. This book has interest in only the finite groups which are sets of permutations (permutation matrices or roots of plus unity if you prefer). Methinks, it might be better to add a fifth axiom saying something about 'must be a set of permutations' to the four traditional group axioms when we define finite groups.

From now on in this book, we will often discard the adjective finite and use only the word group to mean finite groups. In this book, we have no interest in anything other than finite groups.

Summary:

There are four group axioms as listed above. These axioms require an identity element and a unique inverse for every element of the group.

There is only one identity in a finite group.

The inverse of any element in a finite group is unique.

The Cyclic Groups

Chapter 9

The Cyclic Groups

Cyclic groups are denoted by C_n where n is the order of the group. We sometimes see them denoted by $\mathbb{Z}_n$, but we will eschew that notation in this book. Note that the subscript of the cyclic groups is the order of the group; this is different from the symmetric groups where the subscript is the number of objects being permuted.

Cyclic groups are closed sets of permutations. Of course, they are; all finite groups are closed sets of permutations. The closed set of permutations which is a cyclic group is a subset of all permutations of n objects. As such, cyclic groups are all subgroups of some symmetric group, S_n. An example of this is the symmetric group S_3 which has a C_3 subgroup.

The cyclic group permutation matrices:
Choose a positive real integer, say $n = 5$. Draw five $n \times n = 5 \times 5$ square matrices. These matrices will be the adjoint representation of the cyclic group of order $n = 5$, $C_{n=5}$. Now, start with the identity matrix. We know any finite group must have an identity. We have:

$$C_{5-Identity} = \begin{bmatrix} 1 & 0 & 0 & 0 & 0 \\ 0 & 1 & 0 & 0 & 0 \\ 0 & 0 & 1 & 0 & 0 \\ 0 & 0 & 0 & 1 & 0 \\ 0 & 0 & 0 & 0 & 1 \end{bmatrix} \qquad (9.1)$$

Now form a second $n \times n = 5 \times 5$ matrix with the 1's moved one step towards the top right-hand corner and add a 1 into the bottom left-hand corner:

$$\begin{bmatrix} 0 & 1 & 0 & 0 & 0 \\ 0 & 0 & 1 & 0 & 0 \\ 0 & 0 & 0 & 1 & 0 \\ 0 & 0 & 0 & 0 & 1 \\ 1 & 0 & 0 & 0 & 0 \end{bmatrix} \qquad (9.2)$$

Keep repeating this procedure so that the 1's 'dance' across the matrix:

$$\begin{bmatrix} 0 & 0 & 1 & 0 & 0 \\ 0 & 0 & 0 & 1 & 0 \\ 0 & 0 & 0 & 0 & 1 \\ 1 & 0 & 0 & 0 & 0 \\ 0 & 1 & 0 & 0 & 0 \end{bmatrix}, \begin{bmatrix} 0 & 0 & 0 & 1 & 0 \\ 0 & 0 & 0 & 0 & 1 \\ 1 & 0 & 0 & 0 & 0 \\ 0 & 1 & 0 & 0 & 0 \\ 0 & 0 & 1 & 0 & 0 \end{bmatrix}, \begin{bmatrix} 0 & 0 & 0 & 0 & 1 \\ 1 & 0 & 0 & 0 & 0 \\ 0 & 1 & 0 & 0 & 0 \\ 0 & 0 & 1 & 0 & 0 \\ 0 & 0 & 0 & 1 & 0 \end{bmatrix} \qquad (9.3)$$

These five permutation matrices, (9.1) & (9.2) & (9.3), are an adjoint representation of the cyclic group C_5. The same procedure will give an adjoint representation of every cyclic group provided we use matrices of the appropriate size.

There are other adjoint representations of the cyclic groups, but we tend to use the above 'dancing matrices' form of the adjoint representation for the cyclic groups – see (6.11).

The 'dancing matrices' are called circulant matrices[24].

There is a reason that these circulant matrices 'dance' which we will demonstrate with the powers of the C_3 permutation matrices:

$$\begin{bmatrix} 0 & 1 & 0 \\ 0 & 0 & 1 \\ 1 & 0 & 0 \end{bmatrix}^2 = \begin{bmatrix} 0 & 0 & 1 \\ 1 & 0 & 0 \\ 0 & 1 & 0 \end{bmatrix}, \quad \begin{bmatrix} 0 & 1 & 0 \\ 0 & 0 & 1 \\ 1 & 0 & 0 \end{bmatrix}^3 = \begin{bmatrix} 1 & 0 & 0 \\ 0 & 1 & 0 \\ 0 & 0 & 1 \end{bmatrix} \qquad (9.4)$$

---

[24] See: Philip J. Davis : Circulant Matrices.

The Cyclic Groups

Each of the circulant matrices of a particular size is a power of some circulant matrix of the same size – this might be power 1 or power minus 1.

By simply taking successive powers of any circulant matrix, we can generate a cyclic group. This is general to all cyclic groups; in fact, a cyclic group is nothing more than all the powers of a particular element of that group, but we must be cautious. The powers of the 4×4 circulant matrix:

$$C_2 = \left\{ \begin{bmatrix} 0 & 0 & 1 & 0 \\ 0 & 0 & 0 & 1 \\ 1 & 0 & 0 & 0 \\ 0 & 1 & 0 & 0 \end{bmatrix}, \begin{bmatrix} 0 & 0 & 1 & 0 \\ 0 & 0 & 0 & 1 \\ 1 & 0 & 0 & 0 \\ 0 & 1 & 0 & 0 \end{bmatrix}^2 \right\} \quad (9.5)$$

form the group C_2 even though this is a 4×4 matrix. This is the C_2 subgroup of C_4. The powers of each of the matrices:

$$\begin{bmatrix} 0 & 1 & 0 & 0 \\ 0 & 0 & 1 & 0 \\ 0 & 0 & 0 & 1 \\ 1 & 0 & 0 & 0 \end{bmatrix} \& \begin{bmatrix} 0 & 0 & 0 & 1 \\ 1 & 0 & 0 & 0 \\ 0 & 1 & 0 & 0 \\ 0 & 0 & 1 & 0 \end{bmatrix} \quad (9.6)$$

generate the whole C_4 cyclic group of order four.

For any given matrix size, there is always at least one circulant matrix which will generate the whole cyclic group of order equal to the size of the circulant matrix; this element is called a generator of the group. Most often there is more than one such generator matrix. If the size of the circulant matrix is a prime number, then every circulant matrix of that size is a generator of the cyclic group of that order.

The cyclic groups as permutations:
The above permutation matrices, (9.1) to (9.3), have corresponding permutations:

$$\begin{pmatrix} 1 & 2 & 3 & 4 & 5 \\ 1 & 2 & 3 & 4 & 5 \end{pmatrix}, \begin{pmatrix} 1 & 2 & 3 & 4 & 5 \\ 5 & 1 & 2 & 3 & 4 \end{pmatrix}, \begin{pmatrix} 1 & 2 & 3 & 4 & 5 \\ 4 & 5 & 1 & 2 & 3 \end{pmatrix}$$
$$\begin{pmatrix} 1 & 2 & 3 & 4 & 5 \\ 3 & 4 & 5 & 1 & 2 \end{pmatrix}, \begin{pmatrix} 1 & 2 & 3 & 4 & 5 \\ 2 & 3 & 4 & 5 & 1 \end{pmatrix} \tag{9.7}$$

Do you see the pattern?

All cyclic groups can be represented by a similar set of permutations, and such a similar set of permutations is always a cyclic group. Perhaps the reader would like to multiply two such permutations (matrices) together.

How many cyclic groups:
Looking at the above procedure, (9.1) to (9.3), we see that there is a cyclic group for every positive integer. We now have another infinite set of finite groups – the cyclic groups.

Properties of cyclic groups:
Every cyclic finite group is abelian (commutative). The cyclic groups are called cyclic because, well, they are 'kind of' circular. By this we mean, as stated above, that within every cyclic group there is at least one element, usually more than one, the powers of which are every other element of the cyclic group. Technically, there is a generating element, g, in a cyclic group such that $g^0, g^1, g^2, ... g^n$ are all the other elements of the cyclic group.

Note that there are some finite groups, like the dihedral groups, which cannot be wholly generated as the powers of any single element but need two or more generating elements. This 'all elements generated by a single element' is unique to cyclic groups; indeed, it is the defining property of cyclic groups. Let us emphasize that.

A group in which every element of the group can be generated as powers of one element is a cyclic group. Remember, there might be

The Cyclic Groups

more than one element that has this property. Cyclic groups are the only groups that can be generated by a single element.

Cyclic groups in the complex plane:
One way to think of a cyclic group is to picture a unit circle in the complex plane, $\mathbb{C}$. Starting with the real number 1, place a number of points, say $(n-1)$, upon the unit circle in the complex plane. Each point is to be equidistant from each other point as measured along the circumference of the unit circle. These are the roots of plus unity. This set of points, roots, is a representation of the cyclic group C_n; for example, we have C_3:

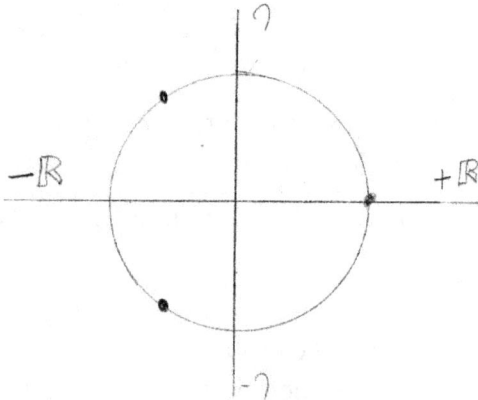

We can list all the cyclic groups as roots of unity:

$$C_1 = +1$$
$$C_2 = \{-1, +1\}$$
$$C_3 = \left\{-\frac{1}{2} + i\frac{\sqrt{3}}{2}, -\frac{1}{2} - i\frac{\sqrt{3}}{2}, +1\right\} \quad (9.8)$$
$$C_4 = \{-i, +i, -1, +1\}$$
$$\ldots$$

The fact that the whole of a cyclic group is generated as powers of a single element is now obvious. For example:

$$i^0 = +1, \quad i^1 = i, \quad i^2 = -1, \quad i^3 = -i, \quad i^4 = +1, \quad i^5 = i,\ldots \quad (9.9)$$

What is also obvious is that not every element of a cyclic group will generate the whole cyclic group; for a start, the identity raised to any power will not generate any other element of the group; further, the minus one element of the C_4 group, (9.9), will not generate the whole group.

Also obvious is the fact that C_4 contains a C_2 subgroup, $\{-1,+1\}$.

The set of elements within a group generated as the powers of a single element is often denoted by $\langle g \rangle$ where g is the element of the group concerned. In any finite group, $\langle g \rangle$ is a subgroup of that finite group. Clearly, since it is generated by a single element, $\langle g \rangle$ is a cyclic subgroup of that finite group.

Not all proper subgroups of a group are cyclic groups; for example, the symmetric group of order twenty-four, S_4, has an order twelve subgroup which is the non-cyclic alternating group A_4.

Adjoint representation of cyclic groups:
The adjoint representations of any finite group are a permutation matrix representation whose sum is the matrix which is all 1's. For cyclic groups, these are the permutation matrices which are the same size as the order of the group. In the case of the order three cyclic group, C_3, the adjoint representation is 3×3 matrices:

$$\begin{bmatrix} 0 & 1 & 0 \\ 0 & 0 & 1 \\ 1 & 0 & 0 \end{bmatrix} + \begin{bmatrix} 0 & 0 & 1 \\ 1 & 0 & 0 \\ 0 & 1 & 0 \end{bmatrix} + \begin{bmatrix} 1 & 0 & 0 \\ 0 & 1 & 0 \\ 0 & 0 & 1 \end{bmatrix} = \begin{bmatrix} 1 & 1 & 1 \\ 1 & 1 & 1 \\ 1 & 1 & 1 \end{bmatrix}_{\text{adjoint}} \quad (9.10)$$

The Cyclic Groups

Fundamental representation of the cyclic groups:
The fundamental representation of a group is the set of permutation matrices of the same size as the number of objects being permuted. In the case of the order three cyclic group, C_3, the fundamental representation is 3×3 matrices because the C_3 cyclic group is a subset of the permutations of three objects. The fundamental representation is also called the defining representation.

We see that, in the case of C_3, the fundamental representation and the adjoint representation are the same representation; this is true for all cyclic groups.

For most groups, the adjoint representation is larger than the fundamental representation; for example, the fundamental representation of the symmetric group S_4 is 4×4 matrices but the adjoint representation of S_4 is 24×24 matrices. Similarly, the fundamental representation of the alternating group A_4, which we are yet to meet, is 4×4 matrices but the adjoint representation of A_4 is 12×12 matrices.

Groups of prime order:
For every prime real number, there is one, and only one, finite group of order equal to that prime real number. This single group is a cyclic group, $\{C_2, C_3, C_5, C_7, ... C_{19}, ... \infty\}$. Since there is a cyclic group of every order, that one group of prime order is, obviously, a cyclic group. In fact, the cyclic groups of prime order are what we call simple finite groups, and they are one of the basic building blocks of groups.

Lagrange's theorem, a preview:
Later we will meet Lagrange's theorem. Lagrange's theorem was presented to the world as a conjecture by Joseph-Louis Lagrange (1736-1813) in 1771. It was partially proven by Carl Friedrich Gauss (1777-

1855) in 1801[25]. It was proven for the symmetric groups by Augustin-Louis Cauchy (1789-1857) in 1844[26] and finally proven for all finite groups by Camille Jordan (1838-1922) in 1861[27].

Lagrange's theorem states that the order of a subgroup divides the order of the group. For example, the order four group C_4 has a C_2 subgroup of order two; 2 divides 4. Another example is the order six cyclic group C_6 which has an order two cyclic subgroup, C_2, and an order three cyclic subgroup, C_3; both 2 & 3 divide 6. Lagrange's theorem means that a group of order n can hold only subgroups whose order divides n.

Thus, by Lagrange's theorem, a group of prime order, G_{prime} has no subgroups other than the identity, which is C_1, and the whole group itself, G_{prime}. Every element of the group G_{prime} other than the identity has powers which are not equal to itself. If the powers of an element were equal to itself, then that element must be the identity, $1^2 = 1$. But the powers of an element are a cyclic group, $\langle g \rangle$. The order of this group, $\langle g \rangle$, must therefore be the same as the order of the prime group, G_{prime}. This G_{prime} must be the cyclic group $\langle g \rangle$. Furthermore, every element of G_{prime} except the identity must similarly generate the whole group.

Suppose there was another group of order equal to G_{prime}, then it too must have an identical subgroup structure to G_{prime}. But if the order and the subgroup structure are the same, then these two groups are the same group.

Thus, for any prime number, we have only one group of prime order and that group is a cyclic group.

---

[25] Gauss : Disquisitiones Artihmeticae 1801
[26] A-L Cauchy : On the products of one or several permutations... : 1844
[27] Camile Jordan 1861 Journal de l'Ecole Polytechnique 22 113-194 pg 166

Single groups of not prime order:
There are orders other than prime order which have only one group. For example, there is only one group of order 85; since there is a cyclic group of every order, this order 85 group is a cyclic group.

The converse of Lagrange's theorem:
The converse of Lagrange's theorem is not true. Just because the order of the group G is divisible by n does not mean that there is a subgroup of order n within that group G. The lowest order group that has this property is the order twelve alternating group, A_4, which has no order six subgroup even though 12 divides by 6.

Abelian groups and cyclic groups:
Although every cyclic group is abelian (commutative), not every abelian group is a cyclic group.

Subgroups of cyclic groups:
Cyclic groups have one and only one subgroup of every order that divides the order of the cyclic group; these subgroups are all cyclic.

Abelian groups in general have a subgroup for every divisor of the group order, but there might be more than one subgroup; for example, the order four group $C_2 \times C_2$ has three C_2 subgroups. Further, when we have subgroups of an abelian group of the same order, these subgroups need not be the same type of group – not isomorphic to each other.

It is only some non-abelian groups which do not have a subgroup of every order that divides the group order.

Finitely generated abelian groups:
Finally, we mention that every finitely generated abelian group (commutative group) is a direct product of cyclic groups. This will not

make sense to the reader at this moment, but this statement ought to be included in the chapter on cyclic groups, and so we have included it.

This statement effectively means that all abelian groups are 'built out of' cyclic groups. That's probably worth repeating.

Summary:
Cyclic groups are generated by one element. Any group generated by one element is a cyclic group. A cyclic group is the set of powers of the one element.

Cyclic groups can be represented by 'dancing' permutation matrices. The fundamental representation of the cyclic group C_n is $n \times n$ matrices. The adjoint representation of the cyclic group C_n is $n \times n$ matrices. The adjoint representation of the cyclic group C_n is the same as the fundamental representation of the cyclic group C_n; this is peculiar to the cyclic groups.

There is a cyclic group of every order.

There is only one finite group of any prime order.

Cyclic groups have only a single cyclic subgroup of every order that divides the order of the cyclic group. There are no other subgroups within a cyclic group.

Abelian groups have at least one subgroup of every order that divides the order of the group.

All abelian groups are 'built out of' cyclic groups. It was worth repeating.

Chapter 10

The Dihedral Groups

We often hear that group theory is the mathematics of symmetry. The whole world seems to be drunk on symmetry, and yet we have tasted only a tiny drop of the symmetry ocean. We need to remember that each type of empty space has its own types of symmetry. A 3-dimensional spinor space, there are four such types of space, that is derived from the C_3 group has 3-dimensional rotational symmetry expressed as the 3-dimensional rotation matrix[28]:

$$\text{3-dim rotation} = \begin{bmatrix} v_A(b,c) & v_B(b,c) & v_C(b,c) \\ v_C(b,c) & v_A(b,c) & v_B(b,c) \\ v_B(b,c) & v_C(b,c) & v_A(b,c) \end{bmatrix} \quad (10.1)$$

wherein the v_i are the 3-dimensional trigonometric functions. This 3-dimensional rotation is very different from the 2-dimensional Euclidean rotation expressed by the rotation matrix:

$$\text{2-dim rotation}_{Euclidian} = \begin{bmatrix} \cos\theta & \sin\theta \\ -\sin\theta & \cos\theta \end{bmatrix} \quad (10.2)$$

Similarly, the reflective symmetries of the 3-dimensional C_3 spaces are different from the 2-dimensional reflective symmetries to which we are accustomed.

Then there are the 4-dimensional spaces derived from the order four finite groups, and then there are the 5-dimensional spaces derived from the order five finite group, and ...

Another example is the rotation matrix of 2-dimensional space-time:

---

[28] See : Dennis Morris : Complex Numbers The Higher Dimensional Forms.

$$\text{2-dim rotation}_{Space-time} = \begin{bmatrix} \cosh \chi & \sinh \chi \\ \sinh \chi & \cosh \chi \end{bmatrix} \quad (10.3)$$

In this space, (10.3), we cannot rotate through 360^0. Thus, we cannot have Euclidean rotational symmetries in 2-dimensional space-time.

Another example is quaternion space which has 'double cover' rotation. In quaternion space, we have to rotate through 720^0 to get back to where we started. Further, quaternion rotation is not commutative.

We see that the type of space is of central importance when we are speaking of symmetry.

Groups and types of space:
We have seen above that every cyclic group can be accommodated within the 2-dimensional Euclidean complex plane, $\mathbb{C}$. The 2-dimensional Euclidean complex numbers are commutative. If follows that the 2-dimensional Euclidean complex plane cannot accommodate non-commutative groups; for example, the 2-dimensional Euclidean complex plane cannot accommodate the symmetric groups. By 'cannot accommodate', we mean that these groups do not exist in 2-dimensional Euclidean space. Similarly, the cyclic groups do not exist, cannot be accommodated, in 2-dimensional space-time.

The order six symmetric group S_3 has a 2-dimensional representation which is:

$$\begin{bmatrix} 1 & 0 \\ 0 & 1 \end{bmatrix}, \begin{bmatrix} -\frac{1}{2} & \frac{\sqrt{3}}{2} \\ -\frac{\sqrt{3}}{2} & -\frac{1}{2} \end{bmatrix}, \begin{bmatrix} -\frac{1}{2} & -\frac{\sqrt{3}}{2} \\ \frac{\sqrt{3}}{2} & -\frac{1}{2} \end{bmatrix}$$

$$\begin{bmatrix} -1 & 0 \\ 0 & 1 \end{bmatrix}, \begin{bmatrix} \frac{1}{2} & \frac{\sqrt{3}}{2} \\ \frac{\sqrt{3}}{2} & -\frac{1}{2} \end{bmatrix}, \begin{bmatrix} \frac{1}{2} & -\frac{\sqrt{3}}{2} \\ -\frac{\sqrt{3}}{2} & -\frac{1}{2} \end{bmatrix} \quad (10.4)$$

The Dihedral Groups

The top row of these, (10.4), are the cyclic group C_3; these three matrices are Euclidean complex numbers – see (4.13). The bottom row of these, (10.4), are not Euclidean complex numbers (the leading diagonal elements are unequal, for a start). We see that the symmetric group S_3 cannot be written as six Euclidean complex numbers. The symmetric group S_3 cannot fit into 2-dimensional Euclidean space.

Euclidean symmetries:
None-the-less, 2-dimensional Euclidean space really exists; it is the complex plane; it is derived from the C_2 finite group. Since 2-dimensional Euclidean space really exists, the symmetries, both rotational and reflective[29], of 2-dimensional Euclidean space really exist. The same is true of all other types of space that really exist, of course.

Furthermore, the 4-dimensional space-time of our universe really exists. This 4-dimensional space-time does not hold 4-dimensional rotations, as observation will verify, but it does hold 2-dimensional Euclidean rotations in three orthogonal 2-dimensional planes. We will use the symmetries of the 2-dimensional Euclidean sub-space of our 4-dimensional space-time to illustrate the dihedral groups.

We point out that the dihedral groups exist independently of our 4-dimensional space-time and independently of 2-dimensional Euclidean space. A dihedral group, like all finite groups, is a closed set of permutations. As such, every dihedral group is a subgroup, a closed subset of permutations, of some symmetric group. Remember, a symmetric group, S_n, is the entire set of permutations of n objects. So, even if it were that no type of space existed, the finite groups, in particular in this chapter, the dihedral groups, would still exist.

---

[29] We have to embed the 2-dimensional complex plane into a higher dimensional space to have reflective symmetries. Since we are not gods, we cannot do this, but nature has done this for us in our 4-dimensional space-time.

Finite Groups – A Simple Introduction

The dihedral groups in Euclidean space:
Above, we looked at the cyclic groups and we drew a picture of a cyclic group as equidistant points on a unit circle in the complex plane. We can think of a cyclic group, like C_3, as being the points at the corners of a regular polygon; this is an equilateral triangle in the case of C_3 and a square in the case of C_4.

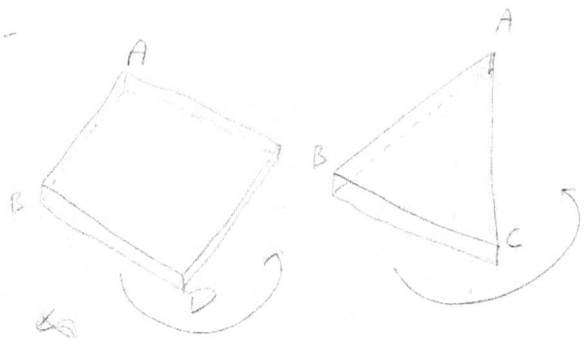

Okay, the art is not great, but the idea is clear. By rotating the equilateral triangle through 120^0 or 240^0, we can permute the vertices of the equilateral triangle. Similarly, by rotating the square through 90^0, 180^0, or 270^0, we can permute the vertices of the square. However, we do not get a full set of possible permutations. There are six permutations of three vertices, but rotating an equilateral triangle gets only three of these permutations. The three rotational permutations of the triangle are the order three C_3 subgroup of the order six symmetric group S_3. The four rotational permutations of the square are four of the 24 possible permutations of four vertices which is the symmetric group S_4. These four permutations are a C_4 subgroup of S_4.

Now, we can flip the polygons over. We need to choose a 'flipping axis' which results in the flipped equilateral triangle 'landing' in exactly the same place as it occupied before it was flipped. In the case of the equilateral triangle, there are three such 'flipping axes' – one through each apex and perpendicular to each side. In the case of the square, there are four such 'flipping axes' – two through and perpendicular to the sides and two through the diagonally opposite corners. In the case of the triangle, flipping in one of these axes swaps the positions of the $\{B,C\}$ vertices:

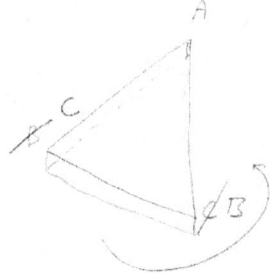

We can now rotate through 120^0 or 240^0 and permute the vertices of the triangle. This time, we get the other three permutations of the S_3 group.

In the case of the square, flipping over and rotating through 90^0, 180^0, and 270^0, will give us another four of the 24 permutations of S_4. In total, rotating and flipping the square has given only eight of the 24 permutations of four vertices. This set of eight permutations is an order eight dihedral group, D_4, which is a subgroup of S_4.

The closed sets of permutations gained from rotating a polygon and flipping a polygon within 3-dimensional Euclidean space are the closed sets of permutations which we call the dihedral groups. There is a dihedral group associated with the triangle, D_3; there is a dihedral group associated with the square, D_4; there is a dihedral group associated with the pentagon, D_5; the hexagon, D_6, etc..

The dihedral group associated with the triangle is denoted as D_3; the three is for the three vertices of a triangle. The dihedral group associated with the square is denoted as D_4; the four is for the four vertices of a square, and so on. The order of a dihedral group is twice the subscript.

Note: We sometimes see the order of the dihedral group written as the subscript rather than seeing half the order written as the subscript. This is a confusion within the notation which does seem to be becoming extinct.

So, a dihedral group has two operations; the first of these is the rotation operation which permutes the corners of the polygon rotationally; the second of these is the 'flip' operation. We can take the view that a dihedral group is generated by two generating operations. We will see that this corresponds to the dihedral groups being generated by two generating elements working together rather than by a single generating element as was the case of the cyclic groups.

Isomorphism between groups:

Clearly, from above, the order six dihedral group, D_3, is the same group as the order six group S_3. It has to be; there are only six permutations of three objects and both the groups $\{S_3, D_3\}$ have all six permutations of three objects. Of course, being a closed set of permutations, each dihedral group is a subgroup of some symmetric group; it is just that, in the case of the order six dihedral group, this subgroup is the whole of the symmetric group.

We have seen above that the order two cyclic group, C_2, is the two permutations of two objects, and so the order two cyclic group is the same group as the order two symmetric group – there is only one set of two permutations of two objects.

When two groups are the same group, we say that they are isomorphic to each other and denote this isomorphism with a $\cong$; we have:

$$C_2 \cong S_2$$
$$D_3 \cong S_3$$
(10.5)

Of course, the order one group is the identity. Every type of group of order one is isomorphic:

$$C_1 \cong S_1 \cong D_1 \cong A_1 \cong ...$$
(10.6)

Truthfully, we should say that two isomorphic groups are the same group and be done with it. The idea of isomorphism is a hang-over from

The Dihedral Groups

days when mathematicians thought of the objects as being the group rather than the permutations of the objects as being the group.

The subgroup structure of the dihedral groups:
Each flip about a 'flipping axis' corresponds to a C_2 subgroup of the appropriate dihedral group. Thus we have three C_2 subgroups in D_3 (the triangle) and five C_2 subgroups in D_5 (the pentagon) and similarly so for all odd sided polygons. We will similarly have four C_2 subgroups in D_4 (the square), but we also have a rotational C_2 subgroup in D_4; this is rotation through 180^0. In general, all even-sided polygons will have this single 180^0 rotational C_2 subgroup.

Thus, all dihedral groups of odd subscript, D_{n-odd} have n C_2 subgroups, and all dihedral groups of even subscript, D_{n-even} have $n+1$ C_2 subgroups.

All dihedral groups, D_n have another cyclic subgroup which is C_n. This corresponds to the unflipped rotations. In general, as well as a single C_2 subgroup, this rotation C_n subgroup will have one other cyclic subgroup corresponding to each divisor of n - we've already counted the C_2 subgroup.

The above is an essential part of the subgroup structure of the dihedral groups, but it is not the entire subgroup structure of the dihedral groups. There are often other complications due to the 'intermixing' of the subgroups; for example, the order twelve dihedral group D_6 has the following subgroups. $7(C_2), C_3, C_6, 3(C_2 \times C_2), 2(D_3)$.

Adjoint representation of the dihedral groups:
Exactly the essence of this subgroup structure is reflected in the adjoint representation of a dihedral group. Actually, exactly this view is reflected in one particular adjoint representation of a dihedral group;

there are many adjoint representations. This particular adjoint representation is of the form:

$$\begin{bmatrix} [C_n]_{Unflipped} & [C_n]_{Flipped} \\ [C_n]_{Flipped} & [C_n]_{Unflipped} \end{bmatrix} \tag{10.7}$$

We will present the adjoint representation as the sum of the separate labelled permutation matrices rather than write out each permutation matrix separately. The individual permutation matrices can be found by simply splitting out the individual variables and setting the label to one. Recall that the C_2 adjoint representation is:

$$\begin{bmatrix} a & b \\ b & a \end{bmatrix} \tag{10.8}$$

Looking at (10.7), we see that that the particular adjoint representation of the dihedral group is basically of the same form as adjoint representation of the order two cyclic group C_2. Into the basic C_2 adjoint representation form, we insert the adjoint representation of the appropriate cyclic group; remember the dancing matrices above, (9.1) to (9.3). We will use C_3 to illustrate this. The adjoint representation of the order three finite cyclic group C_3 is:

$$C_3 = \begin{bmatrix} a & b & c \\ c & a & b \\ b & c & a \end{bmatrix} \tag{10.9}$$

We have, see (10.7):

The Dihedral Groups

$$D_3 \sim \begin{bmatrix} \begin{bmatrix} a & b & c \\ c & a & b \\ b & c & a \end{bmatrix} & [\ ] \\ [\ ] & \begin{bmatrix} a & b & c \\ c & a & b \\ b & c & a \end{bmatrix} \end{bmatrix} = \begin{bmatrix} a & b & c \\ c & a & b \\ b & c & a \\ & & & a & b & c \\ & & & c & a & b \\ & & & b & c & a \end{bmatrix}$$

(10.10)

We then 'flip' the adjoint representation of the appropriate cyclic group. We change the variables to avoid confusion and because the dihedral group is of order twice the order of the appropriate cyclic group:

$$\begin{bmatrix} a & b & c \\ c & a & b \\ b & c & a \end{bmatrix} \xrightarrow{Flip} \begin{bmatrix} c & b & a \\ b & a & c \\ a & c & b \end{bmatrix} \xrightarrow{Re-name} \begin{bmatrix} d & e & f \\ e & f & d \\ f & d & e \end{bmatrix} \quad (10.11)$$

We then place this matrix in the two blank places in (10.10) to produce:

$$D_3 = \begin{bmatrix} a & b & c & d & e & f \\ c & a & b & e & f & d \\ b & c & a & f & d & e \\ d & e & f & a & b & c \\ e & f & d & c & a & b \\ f & d & e & b & c & a \end{bmatrix} \quad (10.12)$$

This, (10.12), is one of the Standard Form Cayley tables of the dihedral group D_3. It is also an algebraic matrix form of this group. We can split out the separate variables to give the separate permutation matrices which are this group.

By the way, we can combine the two groups C_2 & C_3 in other ways; the matrix forms:

$$\begin{bmatrix} [C_2^{a,b}] & [C_2^{c,d}] & [C_2^{e,f}] \\ [C_2^{e,f}] & [C_2^{a,b}] & [C_2^{c,d}] \\ [C_2^{c,d}] & [C_2^{e,f}] & [C_2^{a,b}] \end{bmatrix} \quad \& \quad \begin{bmatrix} [C_3^{a,b,c}] & [C_3^{d,e,f}] \\ [C_3^{d,e,f}] & [C_3^{a,b,c}] \end{bmatrix} \quad (10.13)$$

are both adjoint representations, Standard Form Cayley tables, of the order six cyclic group C_6.

Repeated warning:

The Cayley table gives the product of two elements of the group in the order *top row × left column*. For a commutative group like C_3, we have *top row × left column = left column × top row*, but for a non-commutative group, like D_3 above, (10.12), this is not so. It is a convention that the Cayley table presents the products of two elements of a group in the *top row × left column* order. Looking at (10.12), we have:

$$cd = e \quad \& \quad dc = f \quad (10.14)$$

As multiplying out the separate individual matrices will confirm.

The Cayley table of D_4:

Similarly, using the Standard Form Cayley table of the C_4 group, we can construct the Standard Form Cayley table of the order eight dihedral group, D_4:

The Dihedral Groups

$$D_4 = \begin{bmatrix} a & b & c & d & e & f & g & h \\ d & a & b & c & f & g & h & e \\ c & d & a & b & g & h & e & f \\ b & c & d & a & h & e & f & g \\ e & f & g & h & a & b & c & d \\ f & g & h & e & d & a & b & c \\ g & h & e & f & c & d & a & b \\ h & e & f & g & b & c & d & a \end{bmatrix} \quad (10.15)$$

All the dihedral groups have Standard Form Cayley tables that can be similarly constructed.

A couple of small facts:
There are always only two groups of order $2p$ where p is a prime number. One of these groups is a cyclic group, C_{2p}, and the other is a dihedral group, D_p.

It is possible for a non-abelian group to have subgroups that are all abelian. An example of this is the dihedral group D_3 which has four (cyclic) subgroups all of which are abelian whereas D_3 itself is non-abelian (and non-cyclic).

The dihedral groups as roots of unity:
Seen as roots of unity, the dihedral groups are:

$$\begin{aligned} D_3 &: \quad 1 + 3\sqrt[2]{+1} + 2\sqrt[3]{+1} \\ D_4 &: \quad 1 + 5\sqrt[2]{+1} + 2\sqrt[4]{+1} \\ D_5 &: \quad 1 + 5\sqrt[2]{+1} + 4\sqrt[5]{+1} \\ D_6 &: \quad 1 + 7\sqrt[2]{+1} + 2\sqrt[3]{+1} + 2\sqrt[6]{+1} \\ D_7 &: \quad 1 + 7\sqrt[2]{+1} + 6\sqrt[7]{+1} \end{aligned} \quad (10.16)$$

$$D_8: \quad 1+9\sqrt[2]{+1}+2\sqrt[4]{+1}+4\sqrt[8]{+1}$$
$$D_9: \quad 1+9\sqrt[2]{+1}+2\sqrt[3]{+1}+6\sqrt[9]{+1} \tag{10.17}$$

Another infinite set of finite groups:

We see that there are dihedral groups of every even order. These are groups built from two cyclic groups, C_2 & C_n.

We have not yet written of crossed groups, and so the next two paragraphs might be meaningless to the reader; none-the-less, we feel we should make something clear.

It is generally not the case that the crossed groups[30] $C_2 \times C_n$ are isomorphic to D_n; an example is the order twelve abelian group $C_2 \times C_6$ and the order twelve non-abelian group D_6; these are very different groups.

The case C_2 & C_2 is exceptional in that it is isomorphic to the order four group $C_2 \times C_2$ which we are yet to meet, but since this corresponds geometrically to a polygon with only two vertices, this is not usually considered to be a dihedral group. The Standard Form Cayley table of $C_2 \times C_2$ is of the same form as the dihedral groups in general:

$$C_2 \times C_2 = \begin{bmatrix} a & b & c & d \\ b & a & d & c \\ c & d & a & b \\ d & c & b & a \end{bmatrix} \tag{10.18}$$

The D_6 Standard Form Cayley table:

Since we refer to the order twelve dihedral group later, we present its Standard Form Cayley table here:

---

[30] We will meet the crossed groups later.

The Dihedral Groups

$$D_6 = \begin{bmatrix} a & b & c & d & e & f & g & h & i & j & k & l \\ f & a & b & c & d & e & h & i & j & k & l & g \\ e & f & a & b & c & d & i & j & k & l & g & h \\ d & e & f & a & b & c & j & k & l & g & h & i \\ c & d & e & f & a & b & k & l & g & h & i & j \\ b & c & d & e & f & a & l & g & h & i & j & k \\ g & h & i & j & k & l & a & b & c & d & e & f \\ h & i & j & k & l & g & f & a & b & c & d & e \\ i & j & k & l & g & h & e & f & a & b & c & d \\ j & k & l & g & h & i & d & e & f & a & b & c \\ k & l & g & h & i & j & c & d & e & f & a & b \\ l & g & h & i & j & k & b & c & d & e & f & a \end{bmatrix}$$

(10.19)

We also present two of the permutations corresponding to this adjoint representation of the order twelve dihedral group. These two permutations are illustrative of the nature of the complete set of permutations. The other permutations can be mapped directly from the above (10.19) if needed. Those two permutations are:

$$b = \begin{pmatrix} 1 & 2 & 3 & 4 & 5 & 6 \\ 6 & 1 & 2 & 3 & 4 & 5 \end{pmatrix} \begin{pmatrix} 7 & 8 & 9 & 10 & 11 & 12 \\ 12 & 7 & 8 & 9 & 10 & 11 \end{pmatrix}$$

$$g = \begin{pmatrix} 1 & 7 \\ 7 & 1 \end{pmatrix} \begin{pmatrix} 2 & 12 \\ 12 & 2 \end{pmatrix} \begin{pmatrix} 3 & 11 \\ 11 & 3 \end{pmatrix} \begin{pmatrix} 4 & 10 \\ 10 & 4 \end{pmatrix} \begin{pmatrix} 5 & 9 \\ 9 & 5 \end{pmatrix} \begin{pmatrix} 6 & 8 \\ 8 & 6 \end{pmatrix}$$

(10.20)

We see the C_6 subgroup and we see the six C_2 subgroups.

Summary:
There is a dihedral group of every even order.

Chapter 11

Group Generators

Let us consider the order four cyclic group C_4. We have:

$$C_4 = \begin{bmatrix} a & b & c & d \\ d & a & b & c \\ c & d & a & b \\ b & c & d & a \end{bmatrix} \quad (11.1)$$

This, (11.1), is four permutation matrices which can be separated from each other into the a-permutation matrix, which is the identity, the b-permutation matrix, the c-permutation matrix, and the d-permutation matrix by simply setting these variables to 1.

We see that beginning with the b-permutation matrix, we have:

$$b^1 = b, \quad b^2 = c, \quad b^3 = d, \quad b^4 = a \quad (11.2)$$

We see that this single element has generated the whole group, C_4. We say that the b-permutation is of order four. We have the c-permutation matrix and the d-permutation matrix:

$$\begin{aligned} c^1 &= c, \quad c^2 = a \\ d^1 = d, \quad d^2 &= c, \quad d^3 = b, \quad d^4 = a \end{aligned} \quad (11.3)$$

We see that the c-permutation matrix generates the order two cyclic group C_2 which is a subgroup of C_4. The d-permutation matrix generates the whole group, C_4. We say that the c-permutation is of order two and the d-permutation is of order four.

Group Generators

Inverse elements:

The inverse of the group elements are also powers of the generating element. We have:

$$b^0 = \frac{b^1}{b^1} = \frac{b^2}{b^2} = ... = a \qquad (11.4)$$

We need the inverse elements to form a group, but we automatically have them.

Cyclic groups:

We will be a little repetitive. All cyclic groups are of a similar nature to C_4 in that one of the elements will generate the whole cyclic group. Indeed, this is the defining property of the cyclic groups – the whole group is generated by one element; in any given cyclic group, there are usually many single elements that generate the whole group.

The subgroup structure of cyclic groups is quite simple, but we repeat it anyway. Within a cyclic group of order n, C_n, there is at least one element of order equal to every divisor, m, of the order of the group; for example, the cyclic group C_{12} contains elements of orders 1,2,3,4,& 6. The cyclic groups contain as subgroups one and only one copy of every cyclic group whose order divides the order of the whole group. These subgroups are generated by the elements with the same order as the subgroup.

We see that a single element generates the whole cyclic group. Other types of groups need more than one generating element.

A group with two generators:

Let us consider the dihedral group D_3. We have:

$$D_3 = \begin{bmatrix} a & b & c & d & e & f \\ c & a & b & e & f & d \\ b & c & a & f & d & e \\ d & e & f & a & b & c \\ e & f & d & c & a & b \\ f & d & e & b & c & a \end{bmatrix} \qquad (11.5)$$

The powers of the various permutation matrices of this group, (11.5), are:

$$\begin{aligned}
b^1 &= b, & b^2 &= c, & b^3 &= a \\
c^1 &= c, & c^2 &= b, & c^3 &= a \\
d^1 &= d, & d^2 &= a & & \\
& & e^1 &= e, & e^2 &= a \\
f^1 &= f, & f^2 &= a & &
\end{aligned} \qquad (11.6)$$

We have two elements of order three and three elements of order two, and the identity is of order one, of course. No single element generates the whole dihedral group D_3. If there was such an element, then we would have a cyclic group. All non-cyclic groups are generated by more than one element. The dihedral groups are generated by two elements.

Let us multiply the powers of the b-permutation by the d-permutation; we get:

$$bd = f, \qquad b^2 d = bc = e \qquad (11.7)$$

We already have the powers of the b-permutation generating the c-permutation. We see that the two elements $\{b,d\}$ generate the whole dihedral group.

The same is true of the pairs of elements $\{b,e\}$, $\{b,f\}$, $\{c,d\}$, $\{c,e\}$ and $\{c,f\}$, and so we see that the pair of elements which generates the whole dihedral group is not unique, but there must be two of them.

We note in passing that, since D_3 is the same group as S_3, then S_3 is generated by two elements. In general, S_n is generated by two

Group Generators

generators for $n > 2$. Similarly, $S_n \times S_n$ is generated by two generators. In fact, almost all non-cyclic groups are generated by two generators.

Another example:

In due course, we will meet the group $C_2 \times C_2$. This has adjoint representation, Standard Form Cayley table, algebraic matrix form, call it what you will:

$$C_2 \times C_2 = \begin{bmatrix} a & b & c & d \\ b & a & d & c \\ c & d & a & b \\ d & c & b & a \end{bmatrix} \qquad (11.8)$$

The powers of the different permutation matrices are:

$$b^2 = a, \qquad c^2 = a, \qquad d^2 = a \qquad (11.9)$$

There are three C_2 subgroups in this group, (11.8). We see that no single element generates the whole group. However if we take any two non-identity elements, say $\{b, c\}$, we have $bc = d$. This group is also generated by two generators.

Summary:

Finite groups are generated by a small number, often two, of generating elements called generators. The finite group is the set of all possible products of these generators both with various powers of themselves and with various powers of each other. Of course, a generating element is just a permutation matrix which is just a permutation.

Any group which is generated by a single generating element is a cyclic group.

We see that finite groups are really cyclic groups 'mixed together'. A clear example of this are the dihedral groups discussed above.

For practice, calculate the powers of the permutation matrix:

$$\begin{bmatrix} 1 & 0 & 0 \\ 0 & 0 & 1 \\ 0 & 1 & 0 \end{bmatrix} \quad (11.10)$$

Now, calculate the powers of the permutation matrix:

$$\begin{bmatrix} 0 & 1 & 0 \\ 0 & 0 & 1 \\ 1 & 0 & 0 \end{bmatrix} \quad (11.11)$$

Now, calculate the various products of the various powers of the two permutation matrices, (11.10) & (11.11).

Which finite group has been generated by the two matrices (11.10) & (11.11)?

It is $S_3 = D_3$.

Chapter 12

The Alternating Groups

Every alternating group, A_n, is exactly half of the complete set of permutations of n objects. Of course, the complete set of permutations of n objects is the symmetric group, S_n. Thus, the alternating group, A_n is half of the symmetric group S_n. The order of the symmetric group S_n is $n!$ (n-factorial). Thus, we have it that the order of the alternating group A_n is $\frac{n!}{2}$. Every symmetric group, S_n, has an alternating group, A_n, for a subgroup.

Odd permutations and even permutations:
There are two types of permutation matrices. Half of the $n \times n$ permutation matrices have determinant equal to plus one; we call these the even permutation matrices. The other half of the $n \times n$ permutation matrices have determinant equal to minus one; we call these the odd permutation matrices. Permutation matrices are never singular. Since we have:

$$Det(AB) = Det(A)Det(B) \qquad (12.1)$$

then two even permutation matrices multiplied together are an even permutation matrix. Similarly, two odd permutation matrices multiplied together are an even permutation matrix. It follows that an odd permutation matrix multiplied by an even permutation matrix is an odd permutation matrix.

There are two types of permutations; these are called even permutations and odd permutations. Half of the permutations of any order are odd, and the other half of the permutations of any order are even. We will

Finite Groups – A Simple Introduction

explain odd/even permutations later. For now, we simply state that the alternating group A_n is the set of even permutations of order n.

The alternating group A_3 is three permutations of the six permutations that are S_3, but S_3 has only one subgroup of order three[31] and that is the order three cyclic group, C_3. The order three alternating group is the order three cyclic group, $A_3 \cong C_3$, although the literature often does not acknowledge this; we have:

$$A_3 = \left\{ \begin{pmatrix} 1 & 2 & 3 \\ 1 & 2 & 3 \end{pmatrix} \begin{pmatrix} 1 & 2 & 3 \\ 3 & 1 & 2 \end{pmatrix} \begin{pmatrix} 1 & 2 & 3 \\ 2 & 3 & 1 \end{pmatrix} \right\} \qquad (12.2)$$

$$A_3 = \left\{ \begin{bmatrix} 1 & 0 & 0 \\ 0 & 1 & 0 \\ 0 & 0 & 1 \end{bmatrix}, \begin{bmatrix} 0 & 1 & 0 \\ 0 & 0 & 1 \\ 1 & 0 & 0 \end{bmatrix}, \begin{bmatrix} 0 & 0 & 1 \\ 1 & 0 & 0 \\ 0 & 1 & 0 \end{bmatrix} \right\} \qquad (12.3)$$

We see, as expected, that the alternating group A_3 includes the identity; of course it does; it could not be a group if it did not include the identity.

The alternating group A_4 is of order twelve. The alternating group A_5 is of order sixty.

In the same way that every symmetric group $\{S_{n-1}, S_{n-2}, ..., S_3, S_2, S_1\}$ of order less than n is a proper subgroup of S_n,[32] so every alternating group $\{A_{n-1}, A_{n-2}, ..., A_3, A_2, A_1\}$ of order less than n is a proper subgroup of A_n. This means that we will find the three permutations that are A_3, (12.2), within the alternating group A_4. The three permutations of A_3 include the identity, of course. Clearly, the identity will be within A_4.[33] We have the A_3 permutations within A_4 as:

---

[31] There are also three order two C_2 subgroups in S_3.

[32] Technically, S_n is a subgroup of itself, as are all groups.

[33] This really is clear.

The Alternating Groups

$$A_3 = \left\{ \begin{pmatrix} 1 & 2 & 3 & 4 \\ 1 & 2 & 3 & 4 \end{pmatrix} \begin{pmatrix} 1 & 2 & 3 & 4 \\ 3 & 1 & 2 & 4 \end{pmatrix} \begin{pmatrix} 1 & 2 & 3 & 4 \\ 2 & 3 & 1 & 4 \end{pmatrix} \right\} \quad (12.4)$$

$$A_3 = \left\{ \begin{bmatrix} 1 & 0 & 0 & 0 \\ 0 & 1 & 0 & 0 \\ 0 & 0 & 1 & 0 \\ 0 & 0 & 0 & 1 \end{bmatrix}, \begin{bmatrix} 0 & 1 & 0 & 0 \\ 0 & 0 & 1 & 0 \\ 1 & 0 & 0 & 0 \\ 0 & 0 & 0 & 1 \end{bmatrix}, \begin{bmatrix} 0 & 0 & 1 & 0 \\ 1 & 0 & 0 & 0 \\ 0 & 1 & 0 & 0 \\ 0 & 0 & 0 & 1 \end{bmatrix} \right\} \quad (12.5)$$

We see that holding the fourth element fixed corresponds to having a 1 on the leading diagonal in the bottom right-hand corner of the permutation matrix. A permutation including $4 \to 4$ automatically puts a 1 on the leading diagonal of the corresponding permutation matrix in the fourth column and fourth row.

There are going to be other copies of the A_3 group within the A_4 group in which:

a) the first element is held constant

$$A_3 = \left\{ \begin{bmatrix} 1 & 0 & 0 & 0 \\ 0 & 1 & 0 & 0 \\ 0 & 0 & 1 & 0 \\ 0 & 0 & 0 & 1 \end{bmatrix}, \begin{bmatrix} 1 & 0 & 0 & 0 \\ 0 & 0 & 1 & 0 \\ 0 & 0 & 0 & 1 \\ 0 & 1 & 0 & 0 \end{bmatrix}, \begin{bmatrix} 1 & 0 & 0 & 0 \\ 0 & 0 & 0 & 1 \\ 0 & 1 & 0 & 0 \\ 0 & 0 & 1 & 0 \end{bmatrix} \right\}$$
(12.6)

b) the second element is held constant

$$A_3 = \left\{ \begin{bmatrix} 1 & 0 & 0 & 0 \\ 0 & 1 & 0 & 0 \\ 0 & 0 & 1 & 0 \\ 0 & 0 & 0 & 1 \end{bmatrix}, \begin{bmatrix} 0 & 0 & 1 & 0 \\ 0 & 1 & 0 & 0 \\ 0 & 0 & 0 & 1 \\ 1 & 0 & 0 & 0 \end{bmatrix}, \begin{bmatrix} 0 & 0 & 0 & 1 \\ 0 & 1 & 0 & 0 \\ 1 & 0 & 0 & 0 \\ 0 & 0 & 1 & 0 \end{bmatrix} \right\}$$
(12.7)

c) the third element is held constant

$$A_3 = \left\{ \begin{bmatrix} 1 & 0 & 0 & 0 \\ 0 & 1 & 0 & 0 \\ 0 & 0 & 1 & 0 \\ 0 & 0 & 0 & 1 \end{bmatrix}, \begin{bmatrix} 0 & 0 & 0 & 1 \\ 1 & 0 & 0 & 0 \\ 0 & 0 & 1 & 0 \\ 0 & 1 & 0 & 0 \end{bmatrix}, \begin{bmatrix} 0 & 1 & 0 & 0 \\ 0 & 0 & 0 & 1 \\ 0 & 0 & 1 & 0 \\ 1 & 0 & 0 & 0 \end{bmatrix} \right\} \quad (12.8)$$

The subgroups of A_4 include four C_3 subgroups; $C_3 \cong A_3$.

Above, we have the 4×4 identity permutation matrix and eight other 4×4 permutation matrices of the group A_4. The three other 4×4 permutations and corresponding permutation matrices of the A_4 group are:

$$\begin{pmatrix} 1 & 2 & 3 & 4 \\ 2 & 1 & 4 & 3 \end{pmatrix} \quad \begin{pmatrix} 1 & 2 & 3 & 4 \\ 3 & 4 & 1 & 2 \end{pmatrix} \quad \begin{pmatrix} 1 & 2 & 3 & 4 \\ 4 & 3 & 2 & 1 \end{pmatrix}$$

$$\begin{bmatrix} 0 & 1 & 0 & 0 \\ 1 & 0 & 0 & 0 \\ 0 & 0 & 0 & 1 \\ 0 & 0 & 1 & 0 \end{bmatrix} \quad \begin{bmatrix} 0 & 0 & 1 & 0 \\ 0 & 0 & 0 & 1 \\ 1 & 0 & 0 & 0 \\ 0 & 1 & 0 & 0 \end{bmatrix} \quad \begin{bmatrix} 0 & 0 & 0 & 1 \\ 0 & 0 & 1 & 0 \\ 0 & 1 & 0 & 0 \\ 1 & 0 & 0 & 0 \end{bmatrix} \quad (12.9)$$

Representations of A_4 and S_4 as permutation matrices:

Looking at the 3×3 permutation matrices of A_3, (12.3), we see that if these three permutation matrices are added, we get:

$$\begin{bmatrix} 1 & 0 & 0 \\ 0 & 1 & 0 \\ 0 & 0 & 1 \end{bmatrix} + \begin{bmatrix} 0 & 1 & 0 \\ 0 & 0 & 1 \\ 1 & 0 & 0 \end{bmatrix} + \begin{bmatrix} 0 & 0 & 1 \\ 1 & 0 & 0 \\ 0 & 1 & 0 \end{bmatrix} = \begin{bmatrix} 1 & 1 & 1 \\ 1 & 1 & 1 \\ 1 & 1 & 1 \end{bmatrix} \quad (12.10)$$

We usually use variables to distinguish the different permutation matrices:

The Alternating Groups

$$\begin{bmatrix} a & 0 & 0 \\ 0 & a & 0 \\ 0 & 0 & a \end{bmatrix} + \begin{bmatrix} 0 & b & 0 \\ 0 & 0 & b \\ b & 0 & 0 \end{bmatrix} + \begin{bmatrix} 0 & 0 & c \\ c & 0 & 0 \\ 0 & c & 0 \end{bmatrix} = \begin{bmatrix} a & b & c \\ c & a & b \\ b & c & a \end{bmatrix} \quad (12.11)$$

This is the adjoint representation of A_3; it is also the fundamental representation of A_3. Remember, A_3 is the same group as C_3.

We repeat, for all cyclic groups, C_n, the adjoint representation is $n \times n$ matrices and the fundamental representation is the same $n \times n$ matrices. Thus, for the group $C_3 \cong A_3$, the above, (12.11), is both the adjoint representation and the fundamental representation.

For the group A_4, the fundamental representation is 4×4 matrices because A_4 is a set of permutations of four objects. However, A_4 is of order twelve, therefore the adjoint representation is 12×12 permutation matrices. That adjoint representation includes, of course, the identity permutation matrix.

There are 24 permutations in the symmetric group S_4. The adjoint representation of S_4 is 24×24 permutation matrices, but we can write S_4 with twenty-four 12×12 permutation matrices[34]. Of these twenty-four 12×12 permutation matrices, twelve will be the permutation matrices which are the adjoint representation of the group A_4, which is half of the group S_4.

The 12×12 permutation matrices which are the adjoint representation of A_4 are such that one of them is the identity matrix (a single 1 on every element of the leading diagonal) and the other permutation matrices have no elements on the leading diagonal. Since the sum of the adjoint representation of any group is a square matrix in which every element is a 1, all the other permutation matrices in the adjoint representation must have clear leading diagonals.

---

[34] There are 12! = twelve factorial 12×12 permutation matrices.

Note: The determinant of the exponential of a matrix with zero trace is unity. If we take the exponential of a $n \times n$ permutation matrix with zero trace (all zeros on the leading diagonal), then we get a n-dimensional rotation matrix with n-dimensional trigonometric functions as its elements. The situation is a little more complicated because, other than in dimension two, a single permutation matrix gives only part of the whole rotation matrix[35].

What is the nature of the 12 other 12×12 permutation matrices which form the 'other half' of S_4? There are $12! = $ A Huge Number of 12×12 permutation matrices, and so we have plenty to choose from. The permutation matrices that are the other half of S_4 are such that they too add to form a 12×12 matrix which is all 1's. We demonstrate with S_3 and A_3.

We have seen how the A_3 matrices add above, (12.10). The remaining three S_3 matrices are:

$$\begin{bmatrix} 1 & 0 & 0 \\ 0 & 0 & 1 \\ 0 & 1 & 0 \end{bmatrix} + \begin{bmatrix} 0 & 1 & 0 \\ 1 & 0 & 0 \\ 0 & 0 & 1 \end{bmatrix} + \begin{bmatrix} 0 & 0 & 1 \\ 0 & 1 & 0 \\ 1 & 0 & 0 \end{bmatrix} = \begin{bmatrix} 1 & 1 & 1 \\ 1 & 1 & 1 \\ 1 & 1 & 1 \end{bmatrix} \quad (12.12)$$

Notice that there is a single 1 on the leading diagonal of these three matrices and that these three matrices also sum to an 'all 1s' matrix.

We see that the six 3×3 permutation matrices form two sets and that each of these sets sums to a matrix which is all 1's. The C_3 permutation matrices, (12.10), are the even permutations of three objects. The other three 3×3 permutation matrices, (12.12), are the odd permutations of three objects.

Of the twelve permutation matrices that are the 'other half' of S_4, each has a single 1 on the leading diagonal and each has that single 1 in a different position to the other eleven permutation matrices. Note that

---

[35] See : Dennis Morris : Complex Numbers The Higher Dimensional Forms – 2nd edition.

The Alternating Groups

there are many permutation matrices with a 1 in a particular given position on the leading diagonal.

Even permutations and odd permutations:
Permutations come in two types. The two types are known as even permutations and odd permutations.

If we multiply two odd permutations together, we get an even permutation. If we multiply two even permutations together, we get an even permutation. If we multiply an odd permutation by an even permutation, we get an odd permutation.

In general, an alternating group, A_n, is the set of all even permutations of n objects. This is exactly half of the complete set of permutations of n objects which is S_n.

We will demonstrate this oddness and evenness with the 3×3 representation of S_3, (12.10) & (12.12). The three even permutations of S_3 are the alternating group A_3, (12.10). The product of any two of these three permutation matrices is one of these three permutation matrices – groups are closed sets of permutations.

The other three permutations of the S_3 group are the odd permutations; they are the matrices given in (12.12). However we multiply these three matrices together, we get an even permutation matrix.

If we multiply an odd permutation matrix by an even permutation matrix, we get an odd permutation matrix.

More oddness and evenness:
Above, (6.6), we gave the adjoint representation of the order six symmetric group S_3 as:

$$S_3 = \begin{bmatrix} a & b & c & d & e & f \\ c & a & b & e & f & d \\ b & c & a & f & d & e \\ d & e & f & a & b & c \\ e & f & d & c & a & b \\ f & d & e & b & c & a \end{bmatrix} \quad (12.13)$$

The reader might recall that the even permutations of three objects are the group $A_3 \cong C_3$ and that these appear in the S_3 adjoint matrix as:

$$A_3 = \begin{bmatrix} a & b & c & 0 & 0 & 0 \\ c & a & b & 0 & 0 & 0 \\ b & c & a & 0 & 0 & 0 \\ 0 & 0 & 0 & a & b & c \\ 0 & 0 & 0 & c & a & b \\ 0 & 0 & 0 & b & c & a \end{bmatrix} \quad (12.14)$$

If we multiply this 'even' matrix by itself, we get the product of two even matrices which is of the same 'even' form as above, (12.14). The odd permutations of S_3 appear as:

$$S_3^{odd} = \begin{bmatrix} 0 & 0 & 0 & d & e & f \\ 0 & 0 & 0 & e & f & d \\ 0 & 0 & 0 & f & d & e \\ d & e & f & 0 & 0 & 0 \\ e & f & d & 0 & 0 & 0 \\ f & d & e & 0 & 0 & 0 \end{bmatrix} \quad (12.15)$$

If we multiply this 'odd' matrix by itself, we get the product of two 'odd' matrices which is of the same 'even' form as above, (12.14). Similarly, if we multiply the above 'odd' matrix, (12.15), by the above 'even' matrix, we get an 'odd' matrix.

We have the product of two 'even' matrices is an 'even' matrix; the product of two 'odd' matrices is an 'odd' matrix; the product of an 'even' matrix and an 'odd' matrix is an 'odd' matrix.

This works because the odd/even permutations correspond to the odd/even permutation matrices.

It really is quite astounding how permutations of objects are really matrices in disguise. Perhaps matrices are really permutations of objects in disguise.

The adjoint representation of the symmetric groups:
Above, (12.14), we see the order three cyclic group, $A_3 \cong C_3$, appearing twice on the leading diagonal of the adjoint representation of the symmetric group S_3. In general, the A_n group will appear as the even permutations of the adjoint representation of S_n, and so the algebraic matrix form of the A_n groups will appear in two copies on the leading diagonal of the adjoint representation of S_n.

Summary:
Exactly half of the permutations of $n > 1$ objects are even permutations. The other half of the permutations of $n > 1$ objects are odd permutations.

The even permutations of n objects form a group, A_n, called the alternating group of order $\frac{n!}{2}$.

We have another infinite set of finite groups.

Addendum – determinants:

All permutation matrices are copies of the identity matrix with one or more pairs of columns (could be rows) swapped; for example, swapping columns 2 & 3 gives:

$$\begin{bmatrix} 1 & 0 & 0 \\ 0 & 1 & 0 \\ 0 & 0 & 1 \end{bmatrix} \xrightarrow{\text{Swap column 2 for 3}} \begin{bmatrix} 1 & 0 & 0 \\ 0 & 0 & 1 \\ 0 & 1 & 0 \end{bmatrix} \qquad (12.16)$$

$$\det = +1 \qquad\qquad\qquad \det = -1$$

Every time, we swap a column (row) within any matrix, we change the sign of the determinant of that matrix. We see that an odd permutation matrix is a matrix which differs from the identity matrix by an odd number of column swaps and that an even permutation matrix differs from the identity matrix by an even number of column swaps.

The terms in the determinant of any matrix are the permutations of one element from each row and from each column. There is a sign before each term to be considered. For example, we have:

$$\det\left(\begin{bmatrix} a & b \\ c & d \end{bmatrix}\right) = ad - bc \qquad (12.17)$$

We can write the determinant of any matrix using all the permutation matrices of a size equal to the matrix. Of course, all the permutation matrices of a given size, $n \times n$, are a symmetric group, S_n. We have:

$$\det\left(\begin{bmatrix} a & b & c \\ d & e & f \\ g & h & i \end{bmatrix}\right) =$$

$$\det\left(\begin{bmatrix} 1 & 0 & 0 \\ 0 & 1 & 0 \\ 0 & 0 & 1 \end{bmatrix}\right) \Pr\left(\begin{bmatrix} a & 0 & 0 \\ 0 & e & 0 \\ 0 & 0 & i \end{bmatrix}\right) + \det\left(\begin{bmatrix} 1 & 0 & 0 \\ 0 & 0 & 1 \\ 0 & 1 & 0 \end{bmatrix}\right) \Pr\left(\begin{bmatrix} a & 0 & 0 \\ 0 & 0 & f \\ 0 & h & 0 \end{bmatrix}\right) + \ldots$$

$$= +1.aei - 1.afh + \ldots$$

$$(12.18)$$

We have used the Pr to indicate that we take the product of the three elements in the matrix to which it refers. We calculate the sign before the product, the determinant of the permutation matrix, by counting the number of column swaps by which the permutation matrix differs from the identity (odd or even permutations).

This method will give the determinant of any size of matrix. Since the number of elements in a symmetric group, S_n, is $n!$, then the number of terms in the determinant of a $n \times n$ matrix is $n!$.

Chapter 13

Subgroups

We have now presented several types of finite group to the reader. We have presented the symmetric groups, S_n, the cyclic groups, C_n, the dihedral groups, D_n, and the alternating groups, A_n. There are many more types of groups such as the di-cyclic groups and the crossed groups, also called product groups, which we have not yet presented. We will now change direction a little and look at subgroups.

Subgroups:
Given a subset of a finite group of permutations, we test whether this given subset is a subgroup by checking the subset against the group axioms. We do not need to check multiplicative associativity because this is inherited from the full group of permutations. If the subset satisfies the group axioms, then it is a subgroup of the full group. Every subgroup of a finite group is a finite group of some type in its own right.

Every group contains the two subgroups which are firstly the order one identity element group and secondly the whole group itself.

Subgroups which are neither the identity element nor the whole group are called proper subgroups. Obviously, the order of a proper subgroup is less than the order of the group of which it is a subgroup. There are only a limited number of finite groups of a particular order, and so the proper subgroup must be one of this limited number of the appropriate order.

Within the literature, the reader might often find the group denoted by G and the subgroup denoted by H and the expression $H < G$ indicating that H is a proper subgroup of G.

Subgroups

A few theorems about subgroups:

We might want to know the subgroup structure of a given group. When presented with a large finite group, say of order more than 1000, the subgroup structure of the finite group is not always obvious. Indeed, even for groups of order less than 20, it is tedious to discover every subgroup. The brute force way of mapping the subgroup structure of a finite group is by taking every possible product of every element both with itself and with other elements and examining the sets of these products. There are a lot of such products.

However, we have some theorems and some understanding that will assist in this endeavour. We begin with Lagrange's theorem.

Lagrange's theorem:

We have met Lagrange's theorem earlier in the chapter on cyclic groups. We restate Lagrange's theorem here for the convenience of the reader.

Lagrange's Theorem states that the order of a subgroup of a finite group is always a divisor of the order of the group. An example is the fact that the cyclic group of order twelve, C_{12}, has subgroups with orders $\{2,3,4,6\}$ but it has no subgroups of orders $\{5,7,8,9,10,11\}$.

It is not true that a finite group has a subgroup of every order that is a divisor of the order of the group; for example, the order twelve group A_4 does not have a subgroup of order six.

It is true that every abelian group has at least one subgroup of every order that is a divisor of the order of the group; often, there is more than one subgroup of a particular order.

It is also true that every cyclic group has one, and only one, subgroup of every order that is a divisor of the order of the group. For example, the cyclic group of order twelve, C_{12}, has one of each of the proper subgroups $\{C_2, C_3, C_4, C_6\}$. The order twelve, C_{12} has no proper subgroups other than these $\{C_2, C_3, C_4, C_6\}$.

Cauchy's theorem:

If p is a prime divisor of the order of a finite group, then that finite group has a cyclic subgroup of order p. This subgroup is a cyclic group because it is of prime order p - all groups of prime order are cyclic.

Subgroups of cyclic groups:

With a little thought, we see that every subgroup of a cyclic group is a cyclic group. An example is C_4, see above (13.2) & (13.3) which has a single C_2 subgroup.

The Sylow theorems:

These theorems were discovered and proven by the Norwegian mathematician Peter Ludwig Mejdell Sylow (1832-1918).

If the order of the group G is divisible by a prime p such that p^m is the highest power of p which divides the order of G, then:

1) The group G contains at least one subgroup of order p^m.
2) The number of subgroups of G of order p^m is congruent to $1 \bmod p$ and divides $k = \dfrac{|G|}{p^m}$.

The reader will recall that $|G|$ denotes the order of the group G. The subgroups associated with the Sylow theorems are called Sylow subgroups or Sylow p-subgroups.

For example, the alternating group A_4 has a Sylow 2-subgroup of order $2^2 = 4$; this Sylow 2-subgroup of A_4 is the order four crossed group $C_2 \times C_2$. The order $60 = 2^2 \times 3 \times 5$ alternating group A_5 has, at least, one order two, C_2, subgroup, one order three, C_3, subgroup, one order five,

C_5, subgroup, and one order four, $C_2 \times C_2$, subgroup. It also has the subgroups $\{S_3, A_4, D_{10}\}$ of orders $\{6, 12, 20\}$.

If $p > 2$, the number of Sylow subgroups of order p^m is the same in A_n as it is in S_n. If $n > 5$, A_n and S_n contain the same number of p-Sylow subgroups of even order.

Little snippets:
The Sylow theorems are, with other theorems, very useful in discovering facts about groups of various types. We give a few examples:

i) Any group of order $4n+2$ contains a subgroup of order $2n+1$.

ii) For distinct primes, $p_1, p_2, ... p_n$, if a group of order $p_1, p_2, ... p_n$ is abelian, then this group is a cyclic group.

iii) If the order of a group is $p_1 p_2$ for two distinct primes, all proper subgroups of this group are cyclic groups.

iv) The centre of a group, $C(G)$, is the set of elements which commute with all other elements. The centre of a group is always a subgroup of the group; sometimes, the centre is only the identity.

v) The intersection of two subgroups of a group is also a subgroup of that group.

vi) Every group of order n is a subgroup of S_n.

vii) Any subgroup of an abelian group is also abelian (obvious). Non-abelian groups can have abelian subgroups.

viii) If p, q are prime such that $p > q$ and q is not a divisor of $(p-1)$, then there is only one group of order pq, and it is a cyclic group. Examples are the single group of order $5 \times 17 = 85$ and the single group of order $5 \times 7 = 35$. Note that there are two groups of order $3 \times 7 = 21$ and of order

$5 \times 11 = 55$; in both cases the lesser prime divides the higher prime minus one.

ix) Groups of order p^2 where p is a prime number are of the form C_{p^2} or $C_p \times C_p$; both of these are abelian groups. For example, there are only two groups of order 49, and they are of these forms.

The reader will see that, although we have some understanding of the subgroup structure of the general finite group, this understanding is piecemeal and incomplete. In general, if we wish to know the subgroup structure of a given group, we have to calculate that subgroup structure for each separate group.

Our understanding of groups in general is also piecemeal. If we want to know how many groups there are of a given order and the nature of these groups, most often, we are again reliant on brute force calculation.

Subgroup generators:

For any group, G, the set of elements consisting of the powers of any element of the group G, x, is a cyclic subgroup of the group G. The group of powers of a single generating element, x, is often written as $\langle x \rangle$. We have:

$$\langle x \rangle = \{x^0 = e, x^1, x^2, x^3, ..., x^n\} \qquad (13.1)$$

wherein the identity is denoted by e. The element, x, which generates the group, (13.1), is called the generator of the group, (13.1). Examples of cyclic group generators are the permutation matrices and their powers:

$$\begin{bmatrix} 0 & 1 & 0 & 0 \\ 0 & 0 & 1 & 0 \\ 0 & 0 & 0 & 1 \\ 1 & 0 & 0 & 0 \end{bmatrix} \sim \begin{bmatrix} 0 & 0 & 1 & 0 \\ 0 & 0 & 0 & 1 \\ 1 & 0 & 0 & 0 \\ 0 & 1 & 0 & 0 \end{bmatrix} \sim \begin{bmatrix} 0 & 0 & 0 & 1 \\ 1 & 0 & 0 & 0 \\ 0 & 1 & 0 & 0 \\ 0 & 0 & 1 & 0 \end{bmatrix} \sim \begin{bmatrix} 1 & 0 & 0 & 0 \\ 0 & 1 & 0 & 0 \\ 0 & 0 & 1 & 0 \\ 0 & 0 & 0 & 1 \end{bmatrix}$$

$$(13.2)$$

Subgroups

$$\begin{bmatrix} 0 & 0 & 1 & 0 \\ 0 & 0 & 0 & 1 \\ 1 & 0 & 0 & 0 \\ 0 & 1 & 0 & 0 \end{bmatrix} \sim \begin{bmatrix} 1 & 0 & 0 & 0 \\ 0 & 1 & 0 & 0 \\ 0 & 0 & 1 & 0 \\ 0 & 0 & 0 & 1 \end{bmatrix} \qquad (13.3)$$

The generator matrices are the left-most matrices of (13.2) & (13.3), and the successive powers of these generators are listed across the page. The first of these, (13.2), is the order four cyclic group C_4. The second of these, (13.3), is the order two cyclic group C_2. Note that these two generators are both elements of the order four cyclic group C_4.

The generators, that is each element in a group, are said to have order equal to the power of the generator which gives the identity. For example the order of the generator (element) in (13.2) is four and the order of the element (generator) in (13.3) is two.

It is a consequence of Lagrange's theorem that every element of a group has an order which divides the order of the group. A consequence of this is that any element of the group raised to the order of the group is the identity; we have:

$$x^{|G|} = e \qquad (13.4)$$

Groups of prime order:
If the order of the group, G, is a prime number, then there are no proper subgroups of this group. Recall that all groups of prime order are cyclic groups, and there is one, and only one, cyclic group of any prime order.

Since the order of any subgroup must divide the order of the group, (Lagrange's theorem) if the group is of prime order, then there can be no proper subgroups of that group.

Calculating all the subgroups of a group:
We calculate all the subgroups of a given group by:

a) Using the theorems and snippets above to make our task easier.
b) By considering all possible sets of generator elements taken together. We have seen above, (11.5) to (11.7), how two generators combine to form a group.
c) Using the internet to seek already calculated answers. There are libraries of groups and their subgroup structures available.
d) Find a book listing this information[36]. Such books usually list this information for only the lowest order groups (up to order 32, perhaps).

Summary:

There is no simple and general way of calculating the entire subgroup structure of a given group without laborious perseverance. We have many theorems and snippets to assist us, but it is very easy to 'miss' a subgroup.

In a similar way, it is very easy to 'miss' a group of a given order when listing the finite groups.

Remarkably, as we will see later, there are particular types of groups called the simple finite groups which have been fully catalogued.

List of the subgroups of D_6:

We later frequently refer to the subgroup structure of the order twelve dihedral group D_6. We therefore provide a list of these subgroups for the convenience of the reader. The r element is a generator of the C_6 subgroup; the r is taken from 'rotation'. The s element is the generator of a C_2 subgroup; the s is taken from 'spin' meaning flip. The subgroups of D_6 are:

$$\langle e \rangle = \{e\} \quad : \quad C_1 \qquad (13.5)$$

---

[36] A. D. Thomas & G. V. Wood : Group Tables ISBN 0-906812-04-6 (1980)

Subgroups

$$\langle r \rangle = \langle r^5 \rangle = \{e, r, r^2, r^3, r^4, r^5\} \quad : \quad C_6 \quad (13.6)$$

$$\langle r^2 \rangle = \langle r^4 \rangle = \{e, r^2, r^4\} \quad : \quad C_3 \quad (13.7)$$

$$\langle r^3 \rangle = \{e, r^3\} \quad : \quad C_2 \quad (13.8)$$

$$\langle s \rangle = \{e, s\} \quad : \quad C_2 \quad (13.9)$$

$$\langle rs \rangle = \{e, rs\} \quad : \quad C_2 \quad (13.10)$$

$$\langle r^2 s \rangle = \{e, r^2 s\} \quad : \quad C_2 \quad (13.11)$$

$$\langle r^3 s \rangle = \{e, r^3 s\} \quad : \quad C_2 \quad (13.12)$$

$$\langle r^4 s \rangle = \{e, r^4 s\} \quad : \quad C_2 \quad (13.13)$$

$$\langle r^5 s \rangle = \{e, r^5 s\} \quad : \quad C_2 \quad (13.14)$$

$$\langle r, r^n s \rangle = \{D_6\} \quad : \quad n = 0...5 \quad : \quad D_6 \quad (13.15)$$

$$\langle r^2, s \rangle = \langle r^2, r^2 s \rangle = \langle r^2, r^4 s \rangle = \{e, r^2, r^4, s, r^2 s, r^4 s\} \quad : \quad D_3 \quad (13.16)$$

$$\langle r^2, rs \rangle = \langle r^2, r^3 s \rangle = \langle r^2, r^5 s \rangle = \{e, r^2, r^4, rs, r^3 s, r^5 s\} \quad : \quad D_3 \quad (13.17)$$

$$\langle r^3, rs \rangle = \langle r^3, r^3 s \rangle = \{e, r^3, s, r^3 s\} \quad : \quad C_2 \times C_2 \quad (13.18)$$

$$\langle r^3, rs \rangle = \langle r^3, r^4 s \rangle = \{e, r^3, rs, r^4 s\} \quad : \quad C_2 \times C_2 \quad (13.19)$$

$$\langle r^3, r^2 s \rangle = \langle r^3, r^5 s \rangle = \{e, r^3, r^2 s, r^5 s\} \quad : \quad C_2 \times C_2 \quad (13.20)$$

You see how laborious the calculation of subgroups can be.

Chapter 14

The Dicyclic Groups

The dicyclic groups are denoted by $Q_{4n} : n > 1$. They are non-abelian groups of orders 8, 12, 16,.... The smallest of these groups is the order eight dicyclic group Q_8 which is also known as the quaternion group and is often denoted by simply Q. The subscript is sometimes written as half of the order of the group; the denotation is ambiguous. The Q_{12} group is also denoted as the T group.

The order four dicyclic group is isomorphic to the abelian order four cyclic group, $Q_4 \cong C_4$ and is not considered to be dicyclic.

The dicyclic groups are seen as extensions of the order two cyclic group C_2 by the cyclic group of order $2n$, C_{2n}; hence the name dicyclic.

We sometimes see the denotation Dic_{2n} for these groups.

Generating a dicyclic group:
The dicyclic groups are generated by two elements:

$$x^0 = e, x^1 ... x^{2m-1} \quad \& \quad y, xy ... x^{2m-1}y \qquad (14.1)$$

With the multiplicative relations:

$$x^a x^b = x^{a+b}$$
$$x^a y x^b = x^{a-b} y \qquad (14.2)$$
$$x^a y x^b y = x^{a-b+m}$$

The Dicyclic Groups

The Standard Form Cayley table of Q_8:

Permutations within this group are of the nature of two cyclic groups 'fixed' together; examples of permutations within the order eight Q_8 group are:

$$\begin{pmatrix} 1 & 2 & 3 & 4 & 5 & 6 & 7 & 8 \\ 4 & 1 & 2 & 3 & 6 & 7 & 8 & 5 \end{pmatrix} \qquad \begin{pmatrix} 1 & 2 & 3 & 4 & 5 & 6 & 7 & 8 \\ 7 & 8 & 5 & 6 & 1 & 2 & 3 & 4 \end{pmatrix}$$
$$b \qquad\qquad\qquad\qquad e$$
(14.3)

Of course, these can be picked out of the Standard Form Cayley table, adjoint representation, of the Q_8 group. We have chosen the $\{b,e\}$ variables in (14.3). That Standard Form Q_8 Cayley table is:

$$Q_8 \sim \begin{bmatrix} a & b & c & d & e & f & g & h \\ d & a & b & c & h & e & f & g \\ c & d & a & b & g & h & e & f \\ b & c & d & a & f & g & h & e \\ g & f & e & h & a & d & c & b \\ h & g & f & e & b & a & d & c \\ e & h & g & f & c & b & a & d \\ f & e & h & g & d & c & b & a \end{bmatrix} \qquad (14.4)$$

The connection to the quaternions is given by:

$$a=1,\ b=\hat{i},\ c=-1,\ d=-\hat{i},\ e=j,\ f=k,\ g=-j,\ h=-k$$
(14.5)

Looking at (14.4), the reader might notice two copies of the order four cyclic group C_4 in the top left-hand corner and the bottom right-hand corner. There are similarly two more copies of the C_4 group in the off diagonal corners – hence the name dicyclic.

Finite Groups – A Simple Introduction

Chapter 15

The Crossed Groups – Direct Products

If we list all the 4×4 permutation matrices with zeros on the leading diagonal, we get three sets which, together with the identity, form three copies of the adjoint representation of the C_4 cyclic group; we also get three other such permutation matrices, which, with corresponding permutations written beneath them, are:

$$\begin{bmatrix} 0 & 1 & 0 & 0 \\ 1 & 0 & 0 & 0 \\ 0 & 0 & 0 & 1 \\ 0 & 0 & 1 & 0 \end{bmatrix} \quad \begin{bmatrix} 0 & 0 & 1 & 0 \\ 0 & 0 & 0 & 1 \\ 1 & 0 & 0 & 0 \\ 0 & 1 & 0 & 0 \end{bmatrix} \quad \begin{bmatrix} 0 & 0 & 0 & 1 \\ 0 & 0 & 1 & 0 \\ 0 & 1 & 0 & 0 \\ 1 & 0 & 0 & 0 \end{bmatrix} \quad (15.1)$$

$$\begin{pmatrix} 1 & 2 & 3 & 4 \\ 2 & 1 & 4 & 3 \end{pmatrix} \quad \begin{pmatrix} 1 & 2 & 3 & 4 \\ 3 & 4 & 1 & 2 \end{pmatrix} \quad \begin{pmatrix} 1 & 2 & 3 & 4 \\ 4 & 3 & 2 & 1 \end{pmatrix}$$

Together with the identity, these permutations are the abelian order four $C_2 \times C_2$ group. This group is also known as the Klein group or Klein's group after Felix Klein (1849-1925). It is also known as the 'vier gruppen'.

Looking at the permutations in (15.1), we see that each permutation can be written as two disjoint permutations like:

$$\begin{pmatrix} 1 & 2 & 3 & 4 \\ 3 & 4 & 1 & 2 \end{pmatrix} \equiv \begin{pmatrix} 1 & 3 \\ 3 & 1 \end{pmatrix}\begin{pmatrix} 2 & 4 \\ 4 & 2 \end{pmatrix} \quad (15.2)$$

The $C_2 \times C_2$ group has three C_2 subgroups which are:

$$\begin{pmatrix} 1 & 2 \\ 2 & 1 \end{pmatrix} \& \begin{pmatrix} 1 & 3 \\ 3 & 1 \end{pmatrix} \& \begin{pmatrix} 1 & 4 \\ 4 & 1 \end{pmatrix} \quad (15.3)$$

The Crossed Groups – Direct Products

A different type of representation:

We can also think of $C_2 \times C_2$ as the group:

$$C_2 \times C_2 = \left\{ \begin{pmatrix} +1 \\ +1 \end{pmatrix}, \begin{pmatrix} +1 \\ -1 \end{pmatrix}, \begin{pmatrix} -1 \\ +1 \end{pmatrix}, \begin{pmatrix} -1 \\ -1 \end{pmatrix} \right\} \qquad (15.4)$$

In this representation, we multiply two elements together as we would two independent pairs of real numbers:

$$\begin{pmatrix} +1 \\ -1 \end{pmatrix} \times \begin{pmatrix} -1 \\ -1 \end{pmatrix} = \begin{pmatrix} -1 \\ +1 \end{pmatrix} \qquad (15.5)$$

We can use this type of representation for all groups of the type $C_2 \times C_2 \times ...$ For example, the order eight group $C_2 \times C_2 \times C_2$ is:

$$C_2 \times C_2 \times C_2 = \left\{ \begin{pmatrix} 1 \\ 1 \\ 1 \end{pmatrix}, \begin{pmatrix} -1 \\ 1 \\ 1 \end{pmatrix}, \begin{pmatrix} 1 \\ -1 \\ 1 \end{pmatrix}, \begin{pmatrix} 1 \\ 1 \\ -1 \end{pmatrix}, \begin{pmatrix} -1 \\ -1 \\ 1 \end{pmatrix}, \begin{pmatrix} -1 \\ 1 \\ -1 \end{pmatrix}, \begin{pmatrix} 1 \\ -1 \\ -1 \end{pmatrix}, \begin{pmatrix} -1 \\ -1 \\ -1 \end{pmatrix} \right\}$$

$$(15.6)$$

The remarkableness of $C_2 \times C_2$ - an aside:

This little section is an aside that the reader can ignore if they so choose; it is not part of a standard presentation of finite group theory.

We will see later that the commutative $C_2 \times C_2$ group holds within it non-commutative division algebras like the quaternions. Indeed, it seems that this group is the entire universe except perhaps the strong force. We have shown elsewhere that our 4-dimensional space-time, general relativity, classical electromagnetism, and large parts of QED can be derived from this group[37].[38] There is no other group remotely like this one. Well, the $C_2 \times C_2 \times C_2 \times ...$ groups are like this group in some ways, but these ways are remote.

---

[37] See : Dennis Morris : The Physics of Empty Space
[38] See : Dennis Morris : Upon General Relativity

The Standard Form Cayley table of this group is:

$$C_2 \times C_2 = \begin{bmatrix} a & b & c & d \\ b & a & d & c \\ c & d & a & b \\ d & c & b & a \end{bmatrix} \quad (15.7)$$

The non-commutative algebras of this group, and of the $C_2 \times C_2 \times C_2 \times ...$ groups in general are Clifford algebras[39]. The left-chiral quaternions, which are the Clifford algebra $Cl_{0,2}$, are, in matrix notation:

$$\mathbb{H}_{L\chi} = Cl_{0,2} = \begin{bmatrix} a & b & c & d \\ -b & a & -d & c \\ -c & d & a & -b \\ -d & -c & b & a \end{bmatrix} \quad (15.8)$$

The reader should compare this, (15.8), to the Standard Form Cayley table, (15.7). While we are at it, we might as well compare the complex numbers, $\mathbb{C}$, with the Standard Form Cayley table of the group C_2:

$$\mathbb{C} = \begin{bmatrix} a & b \\ -b & a \end{bmatrix} \qquad C_2 = \begin{bmatrix} a & b \\ b & a \end{bmatrix} \quad (15.9)$$

Crossed groups in general – the direct product:
We can cross any two groups together. We refer to such a crossed product of groups as the direct product of the two groups. There is a procedure for crossing two groups together.

Let the elements of a group G be denoted by g_i and the elements of a group H be denoted by h_i, then the crossed group $G \times H$ is the set of

---

[39] See : Dennis Morris : The Naked Spinor

all possible pairs of elements (g_i, h_j). The groups do not have to be of the same order.

We have:

$$G \times H = H \times G$$
$$A \times (B \times C) = (A \times B) \times C \qquad (15.10)$$

In (15.10), an isomorphism sign is usually used rather than an equals sign, but these two groups are the same group, and so we've used an equals sign.

The order of a crossed group is the product of the orders of the two groups crossed together; in symbols, this is:

$$|G \times H| = |G||H| \qquad (15.11)$$

But I thought finite groups were sets of permutations; how can something like (g_i, h_j) be a permutation? We see that the order of the direct product of two groups is greater than the orders of each of the groups, (15.11). We have seen above, (15.1), that the elements of the group $C_2 \times C_2$ are permutations. Remarkably, a pair of permutations of orders, say $m \& n$, is a permutation of order mn if they are combined together correctly. The correct way to combine them together is to fix them on the diagonals of the larger permutation matrix.

Crossed cyclic groups:
If, and only if, the two crossed groups are both cyclic groups and the highest common factor of the orders of these two groups is 1, then the crossed group will be a cyclic group:

$$C_m \times C_n \cong C_{mn} \quad \text{iff} \quad HCF(m,n) = 1 \qquad (15.12)$$

An example of this is the cyclic group of order six, C_6, which is the same group as $C_2 \times C_3$; however, the orders of the combined cyclic groups must be co-prime to produce a cyclic group; we have $C_2 \times C_4 \neq C_8$ and $C_6 \times C_8$ is not a cyclic group.

There is a converse; if $G \times H$ is a cyclic group, then both $G \& H$ are cyclic groups.

Internal direct product:
If we cross two subgroups, $H \& K$, of a given group, G, together, then that crossed group, $H \times K$, is also a subgroup of the given group G.

However, if both $H \& K$, are normal[40] subgroups of the group, G, and the intersection of $H \& K$ is only the identity, and the products of all elements of $H \& K$ are the group G, then the group G is the same as the direct product of $H \& K$; $G = H \times K$. We say that the finite group $H \times K$ is an internal direct product.

Snippets:
If A is a subgroup of G and B is a subgroup of H, then $A \times B$ is a subgroup of $G \times H$.

The group $G \times H$ is abelian only if both $G \& H$ are abelian.

Crossing permutation matrices:
The cyclic group C_2 is represented by the two 2×2 permutation matrices:

$$\begin{bmatrix} 1 & 0 \\ 0 & 1 \end{bmatrix} \quad \begin{bmatrix} 0 & 1 \\ 1 & 0 \end{bmatrix} \qquad (15.13)$$

How do we cross two such pairs of permutation matrices together?

---

[40] We meet normal subgroups in a later chapter.

The Crossed Groups – Direct Products

Firstly, we know that the order of $C_2 \times C_2$ is four, and so begin by writing a 4×4 representation of C_2. One of the 4×4 permutation matrices will be the identity. There are three possibilities for the other 4×4 permutation matrix; these are given above, (15.1). Then, as a check, we simply add the matrices to form the Standard Form Cayley table, (15.7).

Finite Groups – A Simple Introduction

Chapter 16

Finitely Generated Abelian Groups

We can write any whole positive real number as a product of prime numbers. We would like to be able to do something similar with finite groups. We would like to be able to break any finite group into its simplest parts and express that group as a direct product of those parts.

Every abelian finite group can be written as a direct product of cyclic groups.

Let me just repeat the last sentence. Every abelian finite group, that is every single finite group that is abelian, can be written as a direct product of cyclic groups, that is a set of crossed cyclic groups. Every abelian group can be constructed from only cyclic groups. Abelian groups are no more cyclic groups 'mixed' together.

It follows that we can find every abelian group of a given order by finding all possible direct products of cyclic groups that combine together to produce a group equal to that order. Perhaps we should repeat this sentence also.

We can find every abelian group of any given order.

Finitely generated groups:
A group is finitely generated if it is generated by a finite set of generators. All finite groups are of this nature.

The abelian finitely generated groups can be classified. We do not know how to classify the non-abelian finitely generated groups.

Any finitely generated abelian group is isomorphic to a direct product of cyclic groups, but we have to be a little careful. Direct product is another phrase for cross product. Any finitely generated abelian group is isomorphic to a direct product of cyclic groups of the form:

Finitely Generated Abelian Groups

$$C_a \times C_b \times C_c \times ... \tag{16.1}$$

in which $a > 1$ must be a factor of b and b must be a factor of c etc...

At first sight, the reader might think that this expression, (16.1), might not include a direct product of cyclic groups of the form $C_6 \times C_{10}$ because 6 is not a factor of 10. However:

$$C_6 \times C_{10} = C_2 \times C_3 \times C_{10} = C_2 \times C_{30} \tag{16.2}$$

and $C_2 \times C_{30}$ is included in the expression (16.1) because 2 is a factor of 30. The expression (16.1), does not include $C_3 \times C_4$.

An example of any finitely generated abelian group being isomorphic to a direct product of cyclic groups is that an abelian group of order twelve must be either $C_2 \times C_6$ or C_{12}; in fact, we have both of these groups. An abelian group of order twelve cannot be $C_3 \times C_4$.

Classification of finite abelian groups:
The expression (16.1) and the associated explanation is saying that every abelian finite group can be written in the form given in (16.1). We have to be a little careful, as shown in (16.2), but we can now list every abelian finite group in the form of (16.1). This is sometimes called 'The fundamental theorem of finitely generated abelian groups' – do not forget the abelian bit.

Calculation of the abelian groups of order 360:
The prime decomposition of the number 360 is:

$$2^3 \times 3^2 \times 5^1 = 360 \tag{16.3}$$

We note that the highest prime is 5. We note that there is only one power of 5 in 360. Because 5 is the highest prime, and there is only one power of 5, the direct product presentation of any abelian group of order 360 must be of the form:

$$\ldots \times C_5 \quad (16.4)$$

The power of 3 is 2. We now see that the direct product presentation of any abelian group of order 360 must be one of the two forms:

$$\begin{aligned}\ldots \times C_{3^2 \times 5} &= \ldots \times C_{45} \\ \ldots C_3 \times C_{3 \times 5} &= \ldots C_3 \times C_{15}\end{aligned} \quad (16.5)$$

We have only these two forms because we need the subscript of the leftmost cyclic group to be a factor of the subscript of the cyclic group to its right.

The power of 2 is 3. This gives the possible abelian finite groups of order 360 to be:

$$\begin{aligned} C_{2^3 \times 3^2 \times 5} &= C_{360} \\ C_2 \times C_{2^2 \times 3^2 \times 5} &= C_2 \times C_{180} \\ C_3 \times C_{2^3 \times 3 \times 5} &= C_3 \times C_{120} \\ C_2 \times C_2 \times C_{2 \times 3^2 \times 5} &= C_2 \times C_2 \times C_{90} \\ C_{2 \times 3} \times C_{2^2 \times 3 \times 5} &= C_6 \times C_{60} \\ C_2 \times C_{2 \times 3} \times C_{2 \times 3 \times 5} &= C_2 \times C_6 \times C_{30} \end{aligned} \quad (16.6)$$

We are always looking for the subscript of the leftmost cyclic group to be a factor of the subscript of the cyclic group to its right.

These are the only abelian groups of order 360.

Abelian groups of order p^n:

In general, we can calculate the number of abelian groups of order p^n. For example, we take $n = 5$, and we get the decomposition as:

Finitely Generated Abelian Groups

$$\begin{array}{ccccc} p & p & p & p & p \\ p & p & p & p^2 & \\ p & p^2 & p^2 & & \\ p & p & p^3 & & \\ p^2 & p^3 & & & \\ p & p^4 & & & \\ p^5 & & & & \end{array} \qquad (16.7)$$

We see that for any prime number, p, there are seven abelian groups of order p^5.

The Jordan-Hölder theorem:

The Jordan-Hölder theorem, named after Camille Jordan (1838-1922) and Otto Ludwig Hölder (1859-1937) is a generalisation of 'The fundamental theorem of finitely generated abelian groups' to non-abelian groups. We do not cover this theorem in this introductory book on finite groups.

Chapter 17

Cosets

We consider a group G which has a subgroup H denoted by $H < G$; the elements of the group G are denoted by g_i, and the elements of H are denoted by h_j. Clearly, the h_j are the same as some of the g_i.

Warning:
We repeat a warning given earlier. When we form a product of two elements of a group, we must be careful to keep the order of the elements unchanged. For example, if $\{d, e, f\}$ are three different elements of a finite group, we cannot simply write their product as these three letters in any order. Indeed, we have six different products of these three elements. When forming a product, we have to specify whether we are multiplying on the left or multiplying on the right. We give some examples.

Multiplying d on the left by e is the product ed whereas multiplying d on the right by e is the product de. In general, these two products are not equal. We must be careful to keep the elements in order.

Cosets:
We take a single element of G, say, g_i, and we form the product of every element of the subgroup H by g_i on the left. This gives us a set of products, one product for each element of the subgroup H. These products are each of the form $g_i h_j : j = 1...|H|$. The set of these products is called the left coset of H by g_i. The left coset of H by g_i is a subset (not necessarily subgroup) of the permutations that are the group G. An example is required.

Cosets

Consider the order six dihedral group:

$$D_3 = \begin{bmatrix} a & b & c & d & e & f \\ c & a & b & e & f & d \\ b & c & a & f & d & e \\ d & e & f & a & b & c \\ e & f & d & c & a & b \\ f & d & e & b & c & a \end{bmatrix} \qquad (17.1)$$

The element denoted by a is the identity. This group, (17.1), has an order two cyclic subgroup, $\{a, f\}$. We choose an element of D_3, say b, and we will form the left coset of the subgroup $\{a, f\}$ by b. We have:

$$ba = \begin{bmatrix} 0 & b & 0 & 0 & 0 & 0 \\ 0 & 0 & b & 0 & 0 & 0 \\ b & 0 & 0 & 0 & 0 & 0 \\ 0 & 0 & 0 & 0 & b & 0 \\ 0 & 0 & 0 & 0 & 0 & b \\ 0 & 0 & 0 & b & 0 & 0 \end{bmatrix} \begin{bmatrix} a & 0 & 0 & 0 & 0 & 0 \\ 0 & a & 0 & 0 & 0 & 0 \\ 0 & 0 & a & 0 & 0 & 0 \\ 0 & 0 & 0 & a & 0 & 0 \\ 0 & 0 & 0 & 0 & a & 0 \\ 0 & 0 & 0 & 0 & 0 & a \end{bmatrix} = \begin{bmatrix} 0 & b & 0 & 0 & 0 & 0 \\ 0 & 0 & b & 0 & 0 & 0 \\ b & 0 & 0 & 0 & 0 & 0 \\ 0 & 0 & 0 & 0 & b & 0 \\ 0 & 0 & 0 & 0 & 0 & b \\ 0 & 0 & 0 & b & 0 & 0 \end{bmatrix}$$
(17.2)

$$bf = \begin{bmatrix} 0 & b & 0 & 0 & 0 & 0 \\ 0 & 0 & b & 0 & 0 & 0 \\ b & 0 & 0 & 0 & 0 & 0 \\ 0 & 0 & 0 & 0 & b & 0 \\ 0 & 0 & 0 & 0 & 0 & b \\ 0 & 0 & 0 & b & 0 & 0 \end{bmatrix} \begin{bmatrix} 0 & 0 & 0 & 0 & 0 & f \\ 0 & 0 & 0 & 0 & f & 0 \\ 0 & 0 & 0 & f & 0 & 0 \\ 0 & 0 & f & 0 & 0 & 0 \\ 0 & f & 0 & 0 & 0 & 0 \\ f & 0 & 0 & 0 & 0 & 0 \end{bmatrix} = \begin{bmatrix} 0 & 0 & 0 & 0 & e & 0 \\ 0 & 0 & 0 & e & 0 & 0 \\ 0 & 0 & 0 & 0 & 0 & e \\ 0 & e & 0 & 0 & 0 & 0 \\ e & 0 & 0 & 0 & 0 & 0 \\ 0 & 0 & e & 0 & 0 & 0 \end{bmatrix}$$
(17.3)

We see that the left coset of the subgroup $\{a, f\}$ by b is the two elements $\{b, e\}$.

The right coset of the subgroup $\{a, f\}$ by b is formed by multiplying the elements of the subgroup on the right by b:

$$ab = b \quad \& \quad fb = d \tag{17.4}$$

We see that the right coset of the subgroup $\{a, f\}$ by b is $\{b, d\}$. We see that, in this case, the left coset of the subgroup $\{a, f\}$ by b is not the same as the right coset of the subgroup $\{a, f\}$ by b.

Of course, if the group, G, had been an abelian group, then the right coset would have been equal to the left coset.

Just to remind the reader that the elements of a group are really permutations, we have:

$$b = \begin{pmatrix} 1 & 2 & 3 & 4 & 5 & 6 \\ 3 & 1 & 2 & 6 & 4 & 5 \end{pmatrix} \quad f = \begin{pmatrix} 1 & 2 & 3 & 4 & 5 & 6 \\ 6 & 5 & 4 & 3 & 2 & 1 \end{pmatrix} \tag{17.5}$$

Of course, a is the identity permutation.

Disjoint cosets:
The cosets of a particular subgroup are disjoint. We have a set of n cosets associated with a particular subgroup where n is the number of elements in the group – the order of the group. These cosets differ from each other by the element of the group which was used to generate the coset. These cosets are either equal to each other or are disjoint from each other.

The index of a subgroup:
The number of different disjoint left cosets associated with a subgroup $H < G$ is called the index of the subgroup H in G. The number of left cosets is equal to the number of right cosets.

Cosets

List of the cosets of D_3:

The dihedral group D_3 has three C_2 subgroups, $\{a,d\}$, $\{a,e\}$, & $\{a,f\}$, and one C_3 subgroup $\{a,b,c\}$. We therefore have four cosets by each variable.

The left cosets of each of the four subgroups by a are:

$$\{a,d\} \quad \{a,e\} \quad \{a,f\} \quad \{a,b,c\} \qquad (17.6)$$

Since a is the identity, this is just the subgroups.

The left cosets of each of the four subgroups by b are:

$$\{b,f\} \quad \{b,d\} \quad \{b,e\} \quad \{b,c,a\} \qquad (17.7)$$

The rightmost of these is a subgroup, and so we see that cosets can be subgroups although often cosets are not subgroups.

The left cosets of each of the four subgroups by c are:

$$\{c,e\} \quad \{c,f\} \quad \{c,d\} \quad \{c,a,b\} \qquad (17.8)$$

The left cosets of each of the four subgroups by d are:

$$\{d,a\} \quad \{d,b\} \quad \{d,c\} \quad \{d,e,f\} \qquad (17.9)$$

The left cosets of each of the four subgroups by e are:

$$\{e,c\} \quad \{e,a\} \quad \{e,b\} \quad \{e,f,d\} \qquad (17.10)$$

The left cosets of each of the four subgroups by f are:

$$\{f,b\} \quad \{f,c\} \quad \{f,a\} \quad \{f,d,e\} \qquad (17.11)$$

We will not present the right cosets.

Looking at (17.6) to (17.11), we see that the six cosets of a particular subgroup partition the finite group into disjoint sets. This is general for all groups. By partitioning the set we mean decomposing the set into non-empty subsets no two of which overlap and whose union is the whole set. For example, the order three subgroup cosets shown above are $\{a,b,c\}$ and $\{d,e,f\}$.

Summary:

A coset is a subset of a closed set of permutations.

The left cosets of a subgroup are in one to one correspondence with the right cosets of the subgroup.

The left cosets of a given subgroup, and separately the right cosets of a given subgroup, are either disjoint or are equal.

In this chapter we have described cosets. Doubtless the reader is wondering why we bothered. In a later chapter we will use cosets, but, for now, it is sufficient just to have met them.

Conjugation

Chapter 18

Conjugation

Within a group, some elements are equivalent to each other in a certain way. For example consider the dihedral group D_6; we can think of this as an hexagonal plate in Euclidean space. Looking down from above the hexagonal plate, we rotate the plate clockwise through 60^0. However, this is really the same as rotating the hexagonal plate anti-clockwise through 300^0. To an observer looking at the hexagonal plate from below, our 300^0 anti-clockwise rotation is the same as a 60^0 clockwise rotation. We therefore have either one type of rotation, which is 60^0 clockwise, observed from two different points of view, or we have two types of rotation observed from one point of view. Either way, there is a sense of equivalence between these two rotations.

There is an element in D_6 corresponding to rotation through 60^0 called r, and there is an element in D_6 corresponding to rotation through 300^0 called r^5; within D_6, these two elements are considered to be equivalent in some way. The technical term for this equivalence is conjugacy; we say the two elements $\{r, r^5\}$ are conjugate to each other in D_6. These elements would not necessarily be conjugate to each other in a different group, say D_7. There are no other elements of D_6 that are equivalent to $r \& r^5$, and so we say that these two elements form a conjugacy class.

Finding conjugate elements within a group:
Consider two elements of a group, G, which we will call $\{x, y\}$, and consider an element of G which we call g_i. It will be the case that two of the g_i s are $x \& y$. We say that x is conjugate to y if:

$$gxg^{-1} = y \qquad (18.1)$$

This is the same as (multiply on the right by g):

$$gx = yg \qquad (18.2)$$

We will use the order six dihedral group, D_6, as an example. We will use permutations and $b \,\&\, f$, (17.5), to be $x \,\&\, y$ respectively.

Taking $g = c$:

$$c = \begin{pmatrix} 1 & 2 & 3 & 4 & 5 & 6 \\ 2 & 3 & 1 & 5 & 6 & 4 \end{pmatrix} \qquad (18.3)$$

We have:

$$gxg^{-1} = \begin{pmatrix} 1 & 2 & 3 & 4 & 5 & 6 \\ 2 & 3 & 1 & 5 & 6 & 4 \end{pmatrix} \begin{pmatrix} 1 & 2 & 3 & 4 & 5 & 6 \\ 3 & 1 & 2 & 6 & 4 & 5 \end{pmatrix} \begin{pmatrix} 1 & 2 & 3 & 4 & 5 & 6 \\ 3 & 1 & 2 & 6 & 4 & 5 \end{pmatrix}$$
$$= \begin{pmatrix} 1 & 2 & 3 & 4 & 5 & 6 \\ 3 & 1 & 2 & 6 & 4 & 5 \end{pmatrix} = x$$
$$(18.4)$$

Since the product is not y, then x is not conjugate to y. We chose an abelian subgroup of D_6. With thought, we should have expected this, (18.4), because both g and x are part of an abelian group and, in an abelian group $gxg^{-1} = xgg^{-1} = x$.

Conjugacy within abelian groups and the centre of a group:
Within an abelian group (commutative group), each element is conjugate to only itself. We say that each element is its own conjugacy class.

Within all finite groups, there is a set of elements, which includes the identity, which commute with all other elements. Often, this set of commuting elements is only the identity; in abelian groups this set of

Conjugation

commuting elements is the whole group, but, in some groups, this set of commuting elements is several elements but not the whole group. This set of commuting elements is called the centre of the group and is denoted by $C(G)$ or by $Z(G)$. The centre of a group is always an abelian subgroup of the group, and so each element in the centre of the group is conjugate to only itself.

The centre of both the alternating groups, A_n, and the symmetric groups, S_n, is always only the identity when $n > 3$. The centre of the dihedral groups, D_n, is only the identity if n is odd and is $\left\{e, r^{\frac{n}{2}}\right\}$ when n is even; we have used the letter e to denote the identity.

The generators of D_6:

Clearly, to calculate all the conjugate elements of the order twelve dihedral group involves 144 calculations. Using our wits, we can ease this calculation. The reader might like to think of D_6 as the rotations of a hexagon together with the flipping over of a hexagon.

The group D_6 has a C_6 subgroup generated by a single element; all cyclic groups are generated by single elements. We denote this single element that generates the C_6 subgroup by the letter r. The six elements of this C_6 subgroup then become the ascending powers of r:

$$r, \; r^2, \; r^3, \; r^4, \; r^5, \; r^6 = e = \text{Identity} \qquad (18.5)$$

The other six elements of D_6 are these six elements, (18.5), multiplied by the generator of a C_2 subgroup, which we denote by s. These other six elements of D_6 are thus:

$$rs, \; r^2s, \; r^3s, \; r^4s, \; r^5s, \; r^6s = es = s \qquad (18.6)$$

We see that we have two kinds of elements in D_6. We also have the relations:

$$s^2 = e \quad \& \quad sr = r^5 s \quad \& \quad r^b r^{-b} = e \qquad (18.7)$$

The conjugacy classes of D_6:

To find the elements of D_6 which are conjugate to r^a, we must calculate $g r^a g^{-1}$ for every element, g, of D_6.

Suppose $g = r^b$; the calculation is:

$$r^b r^a r^{-b} = r^a \qquad (18.8)$$

The powers of r are commutative because these are elements of a cyclic group. Thus the element r^a is conjugate to itself, but we have not yet tested all the elements of the group as g.

Suppose $g = s$; the calculation is:

$$s r^a s^{-1} = s r^a s = r^{6-a} s^2 = r^{6-a} \qquad (18.9)$$

We see here that, with $g = s$, the element r^a is conjugate to the element r^{6-a}. We point out that conjugacy exists within a particular group between elements of that particular group. If we were looking at, say, D_8, then the element r^{8-a} would be conjugate to r^a.

Suppose $g = r^b s$; the calculation is:

$$(r^b s) r^a (r^b s)^{-1} = r^b (s r^a s) r^{6-b} = r^b r^{6-a} r^{6-b} = r^{6-a} \qquad (18.10)$$

Thus, the set of elements conjugate to r^a is $\{r^a, r^{6-a}\}$. We say that a set of elements which are conjugate to an element are a conjugacy class.

Thus the set of elements $\{r^a, r^{6-a}\}$ is the conjugacy class of r^a. Of course r^a is an element of the conjugacy class of r^a because all elements are conjugate to themselves.

Conjugation

The complete set of conjugacy classes of D_6

Calculations similar to the above, (18.8) to (18.10) lead to the complete set of conjugacy classes of D_6 which is:

$$\{e\}, \quad \{r,r^5\}, \quad \{r^2,r^4\}, \quad \{r^3\}$$
$$\{s,r^2s,r^4s\}, \quad \{rs,r^3s,r^5s\} \tag{18.11}$$

All elements in a conjugacy class have the same order.

Conjugacy technical:
Conjugacy is an equivalence relation. This means:

a) If x is conjugate to y, then y is conjugate to x. We have:

$$gxg^{-1} = y \Rightarrow x = g^{-1}yg \tag{18.12}$$

b) x is conjugate to itself: $exe^{-1} = x$
c) If x is conjugate to y, and y is conjugate to z, then x is conjugate to z. We have:

$$(g_2g_1)x(g_2g_1)^{-1} = g_2(g_1xg_1^{-1})g_2^{-1} = g_2yg_2^{-1} = z \tag{18.13}$$

In less technical words, elements within a conjugacy class are equivalent to each other in some way. At an abstract level, if we find an equivalence relation, then we know there is some form of equivalence, but we do not know what the equivalence might be.

Permutation matrices and cycle structure:
Consider the permutation:

$$\begin{pmatrix} 1 & 2 & 3 & 4 & 5 & 6 & 7 & 8 & 9 \\ 2 & 1 & 3 & 5 & 6 & 4 & 9 & 7 & 8 \end{pmatrix} \tag{18.14}$$

We can split this permutation into its separate cycles which we call the cycle structure of the permutation; we have:

$$\begin{pmatrix} 1 & 2 & 3 & 4 & 5 & 6 & 7 & 8 & 9 \\ 2 & 1 & 3 & 5 & 6 & 4 & 9 & 7 & 8 \end{pmatrix} = \begin{pmatrix} 1 & 2 \\ 2 & 1 \end{pmatrix}\begin{pmatrix} 3 \\ 3 \end{pmatrix}\begin{pmatrix} 4 & 5 & 6 \\ 5 & 6 & 4 \end{pmatrix}\begin{pmatrix} 7 & 8 & 9 \\ 9 & 7 & 8 \end{pmatrix}$$
(18.15)

We see that we have a 2-cycle, a 1-cycle, and two 3-cycles. The disjoint cycles commute with each other.

We will multiply the permutation (18.15) by itself; that is, we will raise it to power two:

$$\begin{pmatrix} 1 & 2 & 3 & 4 & 5 & 6 & 7 & 8 & 9 \\ 2 & 1 & 3 & 5 & 6 & 4 & 9 & 7 & 8 \end{pmatrix}^2 = \begin{pmatrix} 1 & 2 \\ 1 & 2 \end{pmatrix}\begin{pmatrix} 3 \\ 3 \end{pmatrix}\begin{pmatrix} 4 & 5 & 6 \\ 6 & 4 & 5 \end{pmatrix}\begin{pmatrix} 7 & 8 & 9 \\ 8 & 9 & 7 \end{pmatrix}$$
(18.16)

We see that the 2-cycle has become the order 2 permutation identity. Raised to power three, we have:

$$\begin{pmatrix} 1 & 2 & 3 & 4 & 5 & 6 & 7 & 8 & 9 \\ 2 & 1 & 3 & 5 & 6 & 4 & 9 & 7 & 8 \end{pmatrix}^3 = \begin{pmatrix} 1 & 2 \\ 2 & 1 \end{pmatrix}\begin{pmatrix} 3 \\ 3 \end{pmatrix}\begin{pmatrix} 4 & 5 & 6 \\ 4 & 5 & 6 \end{pmatrix}\begin{pmatrix} 7 & 8 & 9 \\ 7 & 8 & 9 \end{pmatrix}$$
(18.17)

We see that the two 3-cycles have now become the identity but that the 2-cycle is no longer the identity. Raised to power six, the permutation is:

$$\begin{pmatrix} 1 & 2 & 3 & 4 & 5 & 6 & 7 & 8 & 9 \\ 2 & 1 & 3 & 5 & 6 & 4 & 9 & 7 & 8 \end{pmatrix}^6 = \begin{pmatrix} 1 & 2 \\ 1 & 2 \end{pmatrix}\begin{pmatrix} 3 \\ 3 \end{pmatrix}\begin{pmatrix} 4 & 5 & 6 \\ 4 & 5 & 6 \end{pmatrix}\begin{pmatrix} 7 & 8 & 9 \\ 7 & 8 & 9 \end{pmatrix}$$
(18.18)

We see that this permutation is an order 6 permutation; it is an order 6 element of any group of which it is an element.

These cycles clutter around the leading diagonal of the permutation matrix associated with this permutation. That permutation matrix is:

$$P = \begin{bmatrix} 0 & 1 & 0 & 0 & 0 & 0 & 0 & 0 & 0 \\ 1 & 0 & 0 & 0 & 0 & 0 & 0 & 0 & 0 \\ 0 & 0 & 1 & 0 & 0 & 0 & 0 & 0 & 0 \\ 0 & 0 & 0 & 0 & 0 & 1 & 0 & 0 & 0 \\ 0 & 0 & 0 & 1 & 0 & 0 & 0 & 0 & 0 \\ 0 & 0 & 0 & 0 & 1 & 0 & 0 & 0 & 0 \\ 0 & 0 & 0 & 0 & 0 & 0 & 0 & 1 & 0 \\ 0 & 0 & 0 & 0 & 0 & 0 & 0 & 0 & 1 \\ 0 & 0 & 0 & 0 & 0 & 0 & 1 & 0 & 0 \end{bmatrix} \sim \begin{bmatrix} C_2 & \sim & \sim & \sim \\ \sim & C_1 & \sim & \sim \\ \sim & \sim & C_3 & \sim \\ \sim & \sim & \sim & C_3 \end{bmatrix}$$

(18.19)

The separate cycles appear in the permutation matrix as separate elements of groups of lesser order.

The third power of this permutation matrix, (18.19), is 'almost' the identity:

$$P^3 = \begin{bmatrix} 0 & 1 & 0 & 0 & 0 & 0 & 0 & 0 & 0 \\ 1 & 0 & 0 & 0 & 0 & 0 & 0 & 0 & 0 \\ 0 & 0 & 1 & 0 & 0 & 0 & 0 & 0 & 0 \\ 0 & 0 & 0 & 1 & 0 & 0 & 0 & 0 & 0 \\ 0 & 0 & 0 & 0 & 1 & 0 & 0 & 0 & 0 \\ 0 & 0 & 0 & 0 & 0 & 1 & 0 & 0 & 0 \\ 0 & 0 & 0 & 0 & 0 & 0 & 1 & 0 & 0 \\ 0 & 0 & 0 & 0 & 0 & 0 & 0 & 1 & 0 \\ 0 & 0 & 0 & 0 & 0 & 0 & 0 & 0 & 1 \end{bmatrix}$$ (18.20)

This, (18.20), corresponds to (18.17).

The two 1's in the top 2×2 left-hand corner of this permutation matrix, (18.19), oscillate back and forth between being the identity every second power of the matrix. The six 1's in the bottom 6×6 right-hand corner of this permutation matrix, (18.19), oscillate back and forth between being the identity every third power of the matrix. These two oscillations coincide every sixth power of the matrix.

The sixth power of this permutation matrix, (18.19), is the identity. Therefore this permutation matrix is an order six permutation matrix.

We see that, if we are careful, we can read off the order of the permutation from the cycle structure. Above, (18.19), we see the order of this permutation is six.

Conjugacy and cycle structure:
It transpires that all elements of a group that are conjugate to each other have the same cycle structure – same numbers of different lengths of cycles. Thus, all elements of a group that are conjugate to each other have the same order.

The converse is not true. Just because two elements have the same cycle structure does not mean they are conjugate to each other, and not all elements of a group that have the same order are conjugate to each other as is evident within an abelian group.

There is an exception to the converse. All permutations which have the same cycle structure are conjugate to each other in the symmetric groups, S_n. We see below that this is not true in A_n. Whether or not two permutations are conjugate to each other depends upon the group of which they are elements.

Back to conjugacy:
The conjugacy classes of a group partition that group into separate parts. We list the conjugacy classes of the non-abelian group A_4 in the fundamental representation of A_4.

The identity is in a conjugacy class of its own:

$$\left\{ \begin{bmatrix} 1 & 0 & 0 & 0 \\ 0 & 1 & 0 & 0 \\ 0 & 0 & 1 & 0 \\ 0 & 0 & 0 & 1 \end{bmatrix} \right\} \qquad (18.21)$$

Conjugation

The other conjugacy classes of A_4 are:

$$\left\{ \begin{bmatrix} 0 & 0 & 1 & 0 \\ 1 & 0 & 0 & 0 \\ 0 & 1 & 0 & 0 \\ 0 & 0 & 0 & 1 \end{bmatrix}, \begin{bmatrix} 0 & 1 & 0 & 0 \\ 0 & 0 & 0 & 1 \\ 0 & 0 & 1 & 0 \\ 1 & 0 & 0 & 0 \end{bmatrix}, \begin{bmatrix} 0 & 0 & 0 & 1 \\ 0 & 1 & 0 & 0 \\ 1 & 0 & 0 & 0 \\ 0 & 0 & 1 & 0 \end{bmatrix}, \begin{bmatrix} 1 & 0 & 0 & 0 \\ 0 & 0 & 1 & 0 \\ 0 & 0 & 0 & 1 \\ 0 & 1 & 0 & 0 \end{bmatrix} \right\}$$

(18.22)

$$\left\{ \begin{bmatrix} 0 & 1 & 0 & 0 \\ 0 & 0 & 1 & 0 \\ 1 & 0 & 0 & 0 \\ 0 & 0 & 0 & 1 \end{bmatrix}, \begin{bmatrix} 0 & 0 & 0 & 1 \\ 1 & 0 & 0 & 0 \\ 0 & 0 & 1 & 0 \\ 0 & 1 & 0 & 0 \end{bmatrix}, \begin{bmatrix} 0 & 0 & 1 & 0 \\ 0 & 1 & 0 & 0 \\ 0 & 0 & 0 & 1 \\ 1 & 0 & 0 & 0 \end{bmatrix}, \begin{bmatrix} 1 & 0 & 0 & 0 \\ 0 & 0 & 0 & 1 \\ 0 & 1 & 0 & 0 \\ 0 & 0 & 1 & 0 \end{bmatrix} \right\}$$

(18.23)

$$\left\{ \begin{bmatrix} 0 & 1 & 0 & 0 \\ 1 & 0 & 0 & 0 \\ 0 & 0 & 0 & 1 \\ 0 & 0 & 1 & 0 \end{bmatrix}, \begin{bmatrix} 0 & 0 & 1 & 0 \\ 0 & 0 & 0 & 1 \\ 1 & 0 & 0 & 0 \\ 0 & 1 & 0 & 0 \end{bmatrix}, \begin{bmatrix} 0 & 0 & 0 & 1 \\ 0 & 0 & 1 & 0 \\ 0 & 1 & 0 & 0 \\ 1 & 0 & 0 & 0 \end{bmatrix} \right\}$$

(18.24)

We see that each of the twelve elements of A_4 is in a set with others of its kind. We see the four permutations in (18.22) are all of order three, as are the four permutations in (18.23). The four permutations in (18.24) are all of order two. The identity is of order one.

We see that although the permutations in (18.22) have the same cycle structure as the permutations in (18.23), these two sets of permutations are separate conjugacy classes.

Geometrical interpretation of the conjugacy classes of A_4:

We can identify the order twelve alternating group A_4 with the rotational symmetry in 3-dimensional Euclidean space of the apices of a regular tetrahedron. The A_4 set of permutations is the same set as the permutations of the apices of the regular tetrahedron under rotation. The conjugacy classes correspond to particular types of rotations in 3-dimensional Euclidean space.

Given an axis of symmetry through one of the vertices, clockwise rotation through 120^0, corresponds to the four 3-cycle permutations in (18.22). Similarly, anti-clockwise rotation through 120^0, corresponds to the four 3-cycle permutations in (18.23). The identity is rotation though 0^0. The other three permutations, (18.24), are rotations through 180^0 about axes determined by the midpoints of pairs of opposite edges.

The conjugacy classes of S_4:

For the perusal of the reader, we present the conjugacy classes of the order 24 symmetric group S_4. We present them as permutations written in cycle structure. We have:

$$\{Identity\} = \left\{ \begin{pmatrix} 1 & 2 & 3 & 4 \\ 1 & 2 & 3 & 4 \end{pmatrix} \right\} = \left\{ \begin{pmatrix} 1 \\ 1 \end{pmatrix} \begin{pmatrix} 2 \\ 2 \end{pmatrix} \begin{pmatrix} 3 \\ 3 \end{pmatrix} \begin{pmatrix} 4 \\ 4 \end{pmatrix} \right\} \quad (18.25)$$

$$\left\{ \begin{pmatrix} 1 & 2 \\ 2 & 1 \end{pmatrix}\begin{pmatrix} 3 \\ 3 \end{pmatrix}\begin{pmatrix} 4 \\ 4 \end{pmatrix}, \begin{pmatrix} 1 & 3 \\ 3 & 1 \end{pmatrix}\begin{pmatrix} 2 \\ 2 \end{pmatrix}\begin{pmatrix} 4 \\ 4 \end{pmatrix}, \begin{pmatrix} 1 & 4 \\ 4 & 1 \end{pmatrix}\begin{pmatrix} 2 \\ 2 \end{pmatrix}\begin{pmatrix} 3 \\ 3 \end{pmatrix} \right. \\ \left. \begin{pmatrix} 2 & 3 \\ 3 & 2 \end{pmatrix}\begin{pmatrix} 1 \\ 1 \end{pmatrix}\begin{pmatrix} 4 \\ 4 \end{pmatrix}, \begin{pmatrix} 2 & 4 \\ 4 & 2 \end{pmatrix}\begin{pmatrix} 1 \\ 1 \end{pmatrix}\begin{pmatrix} 3 \\ 3 \end{pmatrix}, \begin{pmatrix} 3 & 4 \\ 4 & 3 \end{pmatrix}\begin{pmatrix} 1 \\ 1 \end{pmatrix}\begin{pmatrix} 2 \\ 2 \end{pmatrix} \right\} \quad (18.26)$$

$$\left\{ \begin{pmatrix} 1 & 2 & 3 \\ 2 & 3 & 1 \end{pmatrix}\begin{pmatrix} 4 \\ 4 \end{pmatrix}, \begin{pmatrix} 1 & 2 & 3 \\ 3 & 1 & 2 \end{pmatrix}\begin{pmatrix} 4 \\ 4 \end{pmatrix}, \begin{pmatrix} 1 & 2 & 4 \\ 4 & 1 & 2 \end{pmatrix}\begin{pmatrix} 3 \\ 3 \end{pmatrix} \right. \\ \begin{pmatrix} 1 & 2 & 4 \\ 2 & 4 & 1 \end{pmatrix}\begin{pmatrix} 3 \\ 3 \end{pmatrix}, \begin{pmatrix} 1 & 3 & 4 \\ 3 & 4 & 1 \end{pmatrix}\begin{pmatrix} 2 \\ 2 \end{pmatrix}, \begin{pmatrix} 1 & 3 & 4 \\ 4 & 1 & 3 \end{pmatrix}\begin{pmatrix} 2 \\ 2 \end{pmatrix} \\ \left. \begin{pmatrix} 2 & 3 & 4 \\ 4 & 2 & 3 \end{pmatrix}\begin{pmatrix} 1 \\ 1 \end{pmatrix}, \begin{pmatrix} 2 & 3 & 4 \\ 3 & 4 & 2 \end{pmatrix}\begin{pmatrix} 1 \\ 1 \end{pmatrix} \right\} \quad (18.27)$$

Conjugation

$$\left\{\begin{pmatrix}1&2&3&4\\2&3&4&1\end{pmatrix}, \begin{pmatrix}1&2&3&4\\4&1&2&3\end{pmatrix}, \begin{pmatrix}1&2&3&4\\2&4&1&3\end{pmatrix}\right.$$
$$\left.\begin{pmatrix}1&2&3&4\\3&1&4&2\end{pmatrix}, \begin{pmatrix}1&2&3&4\\3&4&2&1\end{pmatrix}, \begin{pmatrix}1&2&3&4\\4&3&1&2\end{pmatrix}\right\} \quad (18.28)$$

$$\left\{\begin{pmatrix}1&2\\2&1\end{pmatrix}\begin{pmatrix}3&4\\4&3\end{pmatrix}, \begin{pmatrix}1&3\\3&1\end{pmatrix}\begin{pmatrix}2&4\\4&2\end{pmatrix}, \begin{pmatrix}1&4\\4&1\end{pmatrix}\begin{pmatrix}2&3\\3&2\end{pmatrix}\right\} \quad (18.29)$$

The number of conjugacy classes:
There is a part of group theory called representation theory which is concerned with what are called the irreducible representations (irreps) of a group. One of the results of this area of representation theory[41] is that the order of a group is equal to a sum of N square numbers where N is the number of conjugacy classes in a group. For example, there are five conjugacy classes in the order 24 symmetric group S_4 - see (18.25) to (18.29), and we have:

$$24 = 3^2 + 3^2 + 2^2 + 1^2 + 1^2 \quad (18.30)$$

Similarly, there are six conjugacy classes in the order twelve dihedral group D_6 - see (18.11), and we have:

$$12 = 2^2 + 2^2 + 1^2 + 1^2 + 1^2 + 1^2 \quad (18.31)$$

Abelian groups, and only abelian groups, have as many conjugacy classes as is the order of the group, and so the sum of squares is always a sum of several copies of 1^2; for example, the order four cyclic group has:

$$4 = 1^2 + 1^2 + 1^2 + 1^2 \quad (18.32)$$

---

[41] See : Howard Georgi : Lie Algebras in Particle Physics

A non-abelian group can never be like (18.32), and so there are no non-abelian groups of order less than four – a non-abelian group must have at least a single $n^2 : n > 1$, and so must be of order four or above.

The order six symmetric group, S_3, must have three conjugacy classes because it is non-abelian and the only way to write six as a sum of square numbers other than the all 1^2 abelian way, like (18.32), is:

$$6 = 2^2 + 1^2 + 1^2 \qquad (18.33)$$

The actual numbers are the dimensions of the irreducible representations; all the irreducible representations of an abelian group are 1-dimensional. Every group has a trivial irreducible representation of dimension one, and so there will always be at least one 1^2 in the sum of squares. We are now straying too far from the remit of this book.

Summary:
The elements of a finite group can be divided into sets of 'different types' of elements. These sets of 'different types' of elements are called conjugacy classes. The members of these sets have the same cycle structure and are of the same order. Sometimes elements with the same cycle structure, and thus the same order, are members of different conjugacy classes.

The element x is conjugate to the element y if:

$$gxg^{-1} = y \qquad (18.34)$$

For some $g \in G$.

Conjugacy is an equivalence relation. Conjugate elements of a group are equivalent in some way.

Normal Subgroups

Chapter 19

Normal Subgroups

Within finite group theory, there are two types of subgroups. We have what are called normal subgroups, and we have the subgroups which are not normal subgroups.

Normal subgroups have an important role in the classification of what are called simple finite groups. A simple finite group is a group whose only normal subgroups are the identity and the whole group. An example is the cyclic groups of prime order, all of which are simple finite groups.

Normal groups also lead to things called quotient groups or factor groups.

Normal subgroups:
A normal subgroup is a subgroup that is a union of conjugacy classes.

Above, (18.11), we listed the conjugacy classes of the dihedral group of order twelve denoted by D_6. This group, D_6, has a subgroup consisting of the elements $\{e, r^3\}$. Each of these elements is in a conjugacy class by itself, see (18.11). Thus, this subgroup, $\{e, r^3\}$, is a union of conjugacy classes; thus, this subgroup, $\{e, r^3\}$, is a normal subgroup of D_6.

The identity element of any group is a subgroup of that group. The identity element of any group is always in a conjugacy class of its own. Thus, the identity element is always a normal subgroup.

The alternating group, A_n, is always a normal subgroup of the symmetric group S_n. The order six symmetric group S_3 has one normal C_3 subgroup and three C_2 subgroups which are not normal.

The centre, the set of commuting elements, of a group is always a normal subgroup.

In some groups, every subgroup is a normal subgroup; the order eight dicyclic group is an example of this phenomenon. Note that the order eight dicyclic group is a non-abelian group.

Normal subgroups of abelian groups:
Each element of an abelian group is in a conjugacy class of its own. Thus, any subgroup comprised of any set of these elements is a union of conjugacy classes. Thus, any subgroup of an abelian group is a normal subgroup of that group.

Other properties of normal subgroups:
The reader might recall that we discussed cosets a few chapters earlier. The reader might recall that a left coset is the set of all products of the elements of a subgroup, H, multiplied on the left by a given element, b, of the group, G, bH, and that a right coset is the set of all products of the elements of a subgroup, H, multiplied on the right by a given element, b, of the group, G, Hb.

A normal subgroup, H, is a subgroup for which the right cosets are the same as the left cosets for every element of the group, G. Technically, a subgroup, $H < G$ is normal if and only if:

$$xH = Hx \quad \forall \quad x \in G \qquad (19.1)$$

This equality of left cosets and right cosets, (19.1), is sometimes used to define a normal subgroup.

A subgroup $H < G$ is is normal if:

Normal Subgroups

$$ghg^{-1} \in H \quad \forall h \in H \quad \& \quad \forall g \in G \qquad (19.2)$$

This property, (19.2), is also sometimes used to define a normal subgroup. The property (19.2) is a useful way of showing that a given subgroup is a normal subgroup without having to calculate the conjugacy classes of the group.

The expression (19.2) is really just the same as (19.1). Multiplying (19.2) on the right by g gives:

$$gH = Hg \qquad (19.3)$$

If you prefer, a normal subgroup is a subgroup which is invariant under conjugation by all members of the group of which it is a subgroup.

Normal subgroups of normal subgroups:
Consider a group K which is a normal subgroup of a group H which is itself a normal subgroup of a group G, $K < H < G$.

The group K is clearly a subgroup of G, but, surprisingly, it is not necessarily the case that K is a normal subgroup of G even though K is a normal subgroup of H and H is a normal subgroup of G. Technically, normality is not transitive.

However, if K is a normal subgroup of G, then H must also be a normal subgroup of G.

An example of this non-transitivity of normality is the order eight dihedral group D_4. In contradistinction, within the order twelve dicyclic group, Q_{12}, normality is transitive.

However, if H is a cyclic normal subgroup of G, then any subgroup of H is a normal subgroup of G.

The commutator subgroup:

Let two elements of the group G be denoted by $\{g,h\}$. The commutator of these two elements is written as $[g,h]$ and is defined as:

$$[g,h] = g^{-1}h^{-1}gh \tag{19.4}$$

If the two elements $\{g,h\}$ commute, $gh = hg$, then we have it that the set of commutators is only the identity:

$$[g,h] = g^{-1}h^{-1}gh = g^{-1}gh^{-1}h = e = Identity \tag{19.5}$$

wherein we have used the letter e to denote the identity.

Within a group, the product $g^{-1}h^{-1}gh$ is an element of the group. If the group is abelian, then the product $g^{-1}h^{-1}gh$ is the identity for all elements of the group $\{g,h\} \in G$. If the group is non-abelian, then there will be some other elements besides the identity which are the commutators of pairs of elements of the group.

The set of commutators of the group G is not necessarily a subgroup of the group G, but the set of commutators of the group G together with all the products of these commutators is a subgroup of the group G, and it is called the commutator subgroup. If a group, G, is such that the set of commutators is the whole group, $[g_i, g_j] = G \ \forall \ g_i \in G$, then the group G is said to be a perfect group. In a sense, a perfect group is the opposite of an abelian group.

The order eight finite group $C_2 \times C_2 \times C_2$ has two commutative elements and six non-commutative elements; this is not a perfect group. The order sixteen finite group $C_2 \times C_2 \times C_2 \times C_2$ has only one commutative element, the identity, and fifteen non-commutative elements; this is a perfect group.

The commutator subgroup is always a normal subgroup of the group which holds it. For abelian groups, the commutator subgroup is only the identity.

Normal Subgroups

The commutator subgroup of a group is thought of as being a measure of the 'amount of non-commutativity' in the group. The least non-commutative groups are the abelian groups whose commutator subgroup is of order one – just the identity.

The commutator subgroup of the symmetric group S_n is the alternating group A_n. The commutator subgroup of the order twelve alternating group A_n is the Klein group $C_2 \times C_2$.

Notation:
Within the literature, the expression $H \triangleleft G$ is often used to signify that H is a normal subgroup of G rather than $H < G$ which signifies only that H is a proper subgroup of G.

Summary:
There are two types of subgroups known as normal subgroups and as not normal subgroups.

A subgroup is normal if it is a union of conjugacy classes.

A subgroup is a normal subgroup of a group if the right cosets formed by multiplying this subgroup on the right are the same as the left cosets formed by multiplying this subgroup on the left, $gH = Hg$. Above, (17.4), we saw that the subgroup $\{a, f\}$ of D_3 has right cosets that are different from its left cosets. Thus, the subgroup $\{a, f\}$ of D_3 is not a normal subgroup of D_3.

Finite Groups – A Simple Introduction

Chapter 20

Quotient Groups

A finite group is a closed set of permutations. The finite group is not the set of objects being permuted. However, until recently, mathematicians did think of a finite group as being a set of objects like rotations, coloured balls, etc. This led to the idea that two different sets of objects are isomorphic if the permutations of those two sets of objects are the same set of permutations. This idea still hangs over, are we often see the phrase 'isomorphic' used to describe the relationship between two copies of the same group. Many group theorists today will say things like, "The rotations of an equilateral triangle are the group C_3" even though they actually know better.

Part of that hang over is the idea that cosets, which are sets of permutations, form a group. Such groups of cosets are called quotient groups.

Multiplying two sets of permutations together:
If $X = \{x_i\}$ is a subset of the elements of a group, G; that is, X is a subset of the permutations that form the group G, and $Y = \{y_j\}$ is a subset of the elements of a group, G, then we can multiply these elements together as the set of products $x_i y_j$. Of course, these products will be elements of the group G, but they will be a particular set of those group elements.

We have an example; consider the two sets of permutations:

$$\left\{ \begin{pmatrix} 1 & 2 & 3 \\ 3 & 1 & 2 \end{pmatrix}, \begin{pmatrix} 1 & 2 & 3 \\ 3 & 2 & 1 \end{pmatrix} \right\} \ \& \ \left\{ \begin{pmatrix} 1 & 2 & 3 \\ 2 & 3 & 1 \end{pmatrix}, \begin{pmatrix} 1 & 2 & 3 \\ 1 & 3 & 2 \end{pmatrix} \right\} \quad (20.1)$$

The product of these two sets of permutations are the four permutations:

Quotient Groups

$$\begin{pmatrix} 1 & 2 & 3 \\ 3 & 1 & 2 \end{pmatrix}\begin{pmatrix} 1 & 2 & 3 \\ 2 & 3 & 1 \end{pmatrix} = \begin{pmatrix} 1 & 2 & 3 \\ 1 & 2 & 3 \end{pmatrix}$$

$$\begin{pmatrix} 1 & 2 & 3 \\ 3 & 1 & 2 \end{pmatrix}\begin{pmatrix} 1 & 2 & 3 \\ 1 & 3 & 2 \end{pmatrix} = \begin{pmatrix} 1 & 2 & 3 \\ 3 & 2 & 1 \end{pmatrix}$$

$$\begin{pmatrix} 1 & 2 & 3 \\ 3 & 2 & 1 \end{pmatrix}\begin{pmatrix} 1 & 2 & 3 \\ 2 & 3 & 1 \end{pmatrix} = \begin{pmatrix} 1 & 2 & 3 \\ 2 & 1 & 3 \end{pmatrix}$$

$$\begin{pmatrix} 1 & 2 & 3 \\ 3 & 2 & 1 \end{pmatrix}\begin{pmatrix} 1 & 2 & 3 \\ 1 & 3 & 2 \end{pmatrix} = \begin{pmatrix} 1 & 2 & 3 \\ 3 & 1 & 2 \end{pmatrix}$$

(20.2)

The subsets of the group G can be cosets of a given subgroup.

If a subgroup, $H < G$ is a normal subgroup of G, then the set of all left cosets of H, which are the same as the right cosets of H because H is normal, form a group of cosets, called the quotient group, under multiplication of their elements as $x_i y_j$ described above, (20.2).

It is important to realise that the elements of the quotient group are not single permutations but are sets of permutations which are the cosets of some subgroup. The elements of the quotient group are cosets, sets of permutations, not permutations.

If H is not a normal subgroup, then the set of all left cosets of H, do not form a group under multiplication of their elements as $x_i y_j$ described above.

The quotient of the cosets of the normal subgroup $H \triangleleft G$ is denoted by G/H and is often called the factor group.

An example of a quotient group:
The order twelve dihedral group, D_6 has conjugacy classes listed in (18.11). Of these conjugacy classes two are $\{e\}$ and $\{r^3\}$. The two elements $\{e, r^3\}$ of D_6 are a C_2 subgroup of D_6. Because this subgroup

is a union of conjugacy classes, this subgroup, $\{e, r^3\}$, is a normal subgroup of D_6. We will denote this C_2 subgroup of D_6 as H.

There are six distinct cosets of H; they are:

$$eH = \{e, r^3\}, \quad rH = \{r, r^4\}, \quad r^2 H = \{r^2, r^5\}$$
$$sH = \{s, r^3 s\}, \quad rsH = \{rs, r^4 s\}, \quad r^2 sH = \{r^2 s, r^5 s\} \quad (20.3)$$

These six cosets of D_6 are the factor group (quotient group) $D_6 / H = D_6 / C_2$. There are two order six groups, C_6 & $S_3 = D_3$. By multiplying the cosets, (20.3), together as (20.2) and examining the products, we see that we have the order six symmetric group. Thus we can write:

$$D_6 / C_2 \cong S_3 \quad (20.4)$$

We have chosen to use the isomorphism sign here rather than the equals sign. Remember, S_3 is six permutations but D_6 / C_2 is six sets of permutations – six cosets.

The $eH = (e, r^3)$ coset is the identity of the group; for example:

$$\{e, r^3\} \times \{r, r^4\} = \{r, r^4, r^4, r^7 = r\}$$
$$\{e, r^3\} \times \{rs, r^4 s\} = \{rs, r^4 s, r^4 s, r^7 s = rs\} \quad (20.5)$$

Quotient groups of cyclic groups:
The quotient groups of cyclic groups are themselves cyclic.

Chapter 21

Simple Finite Groups

A simple finite group is a group whose order is greater than one and whose only normal subgroups are the identity, $\{e\}$, and the whole group. An example is any of the cyclic groups of prime order, C_p; another example is the alternating groups, $A_n : n \geq 5$ of order 60 and above.

Jordan-Hölden theorem:
Simple finite groups can be seen as the basic building blocks of all finite groups. The construction is by direct products. Simple finite groups are to finite groups 'almost' what prime numbers are to the whole real numbers. The proof of this is called the Jordan-Hölden theorem.

The 'almost' in the previous paragraph is a significant difference between prime numbers and simple finite groups. Within the real numbers, any product of prime numbers, $p_1 p_2 ... p_n$ is a unique real number; such a product of prime numbers is not two or more real numbers; it is only one real number. However, a direct product of simple groups can be more than one type of group. We say there might be many non-isomorphic groups with the same composition series. This means the product of many simple groups is not necessarily a unique particular group.

The classification of the finite simple groups:
Remarkably, the simple finite groups have all been classified. The whole task took more than a century, starting with Evarist Galois (1811-1832) in 1832, but the majority of it was done between 1955 and 2004. The proof that this classification is complete consists of tens of

thousands of pages spread across over a hundred journals and authored by over a hundred different mathematicians. The story of how the classification of the simple finite groups was established is a heroic epic of mathematical history[42]. The classification of the finite simple groups is known as 'The Classification Theorem'.

The proof of the classification theorem was prematurely announced in 1983 by Daniel E. Gorenstein (1923-1992), but it was soon seen to be incomplete. The final proof was presented to the world in 2004 by Michael Aschbacher[43] (1944-).

It is believed that the proof of the classification theorem, being many tens of thousands of pages long, could be cut down to something more manageable if we understood this area of mathematics more deeply. As I write, the mathematicians Richard Neil Lyons (1945-) and Ronald Solomon (1948-), building upon the work and contributions of Gorenstein, are slowly publishing exactly such a simplification.

The simple finite groups:
The simple finite groups are of the forms:

a) The cyclic groups of prime order, C_p.
b) The alternating groups, A_n, of degree more than four.
c) The simple groups of Lie type.
d) One of the 26[44] sporadic groups. These are also called the exceptional groups.

---

[42] See: Mark Ronan : Symmetry and the Monster.
[43] Michael Aschbacher (2004) The status of the classification of the finite simple groups. Notices of the American Mathematical Society 51 (7) 736-740
[44] We sometimes see the sporadic groups numbered as 27 rather than as 26. The extra 'sporadic' is the order 17,971,200 Tits group which is considered not to be a sporadic group by most mathematicians.

The 26 sporadic simple finite groups:

It is quite shocking that, as well as having three infinite sets of simple finite groups, we also have 26 odd balls that do not fit into any classification scheme. Each of these sporadic groups has its own distinct properties. The quest to find these 26 sporadic groups and to prove that there are only 26 of them spans many individuals and many decades[45]. The largest of these sporadic groups is called the monster. The order of the monster is exactly known, but we do not give the exact number. The order of the monster is approximately 8×10^{53} - hence the name monster.

The monster moonshine theorem:

Remarkably, the monster group is connected to 32-dimensional string theory. The connection is called 'The Monster Moonshine theorem'. This theorem was proved by Richard Bocherds (1959-) in 1992. He was awarded the Fields medal in 1998 for producing this proof[46].

The Atlas of Finite Groups:

The classification of the finite simple groups is published in:

Atlas of Finite Groups

By

J. H. Conway

R. T. Curtis

S. P. Norton

R. A. Parker

R. A. Wilson

---

[45] See: Mark Ronan : Symmetry and the Monster.
[46] Bocherds Richard (1992) Monstrous Moonshine and Monstrous Lie superalgebras Invent. Math. 109 405-444

The reader might notice that the names of the authors. These names all have two initials. These names all have surnames of the same length, and the surnames all have vowels in the second and fifth places.

The mathematician Marcus du Sautoy (1965-) tells the story that, upon graduation, he applied to be one of this team, the authors above, who were researching for the Atlas of Finite Groups. He was accepted on the condition that he changed his name to fit with the names of the other authors. He declined thinking these authors to be too eccentric.

Chapter 21

From Finite Group to Division Algebra

Finite groups are very interesting, and there are many technological applications of finite group theory. None-the-less, we leave the details of the technological applications to others, and we proceed in only one application of finite group theory. In this and the next chapter, we will present to the reader an outline of how our 4-dimensional space-time emerges from the finite groups. We present only a skeleton outline; more details can be found in the books listed at the end of this book.

Division algebras (types of complex numbers):
Take the adjoint representation of the finite group C_2. The adjoint representation is the Standard Form Cayley table. The adjoint representation of C_2 is the matrix:

$$C_2 \sim \begin{bmatrix} a & b \\ b & a \end{bmatrix} \qquad (22.1)$$

Let the different variables within this matrix be real numbers and take the exponential of this matrix:

$$\exp\left(\begin{bmatrix} a & b \\ b & a \end{bmatrix}\right) = \begin{bmatrix} e^a & 0 \\ 0 & e^a \end{bmatrix}\begin{bmatrix} \cosh b & \sinh b \\ \sinh b & \cosh b \end{bmatrix} \qquad (22.2)$$

Voila! We have 2-dimensional space-time. This is the transformation of special relativity:

$$\begin{bmatrix} \cosh b & \sinh b \\ \sinh b & \cosh b \end{bmatrix} = \begin{bmatrix} \gamma & v\gamma \\ v\gamma & \gamma \end{bmatrix} \qquad (22.3)$$

$$\gamma = \frac{1}{\sqrt{1-\frac{v^2}{c^2}}} \qquad (22.4)$$

The above, (22.2), is the polar form of the hyperbolic complex numbers first discovered by James Cockle (1819-1995) in 1848. The hyperbolic complex numbers are a division algebra of the same algebraic status as the Euclidean complex numbers. Note that the polar form is the division algebra; the Cartesian form has an element with zero determinant and thus no inverse.

Putting $d = e^a$ and putting the polar form of the hyperbolic complex numbers equal to the Cartesian form of the hyperbolic complex numbers and taking the determinants of both sides gives the distance function of 2-dimensional space-time:

$$\det\left(\begin{bmatrix} d & 0 \\ 0 & d \end{bmatrix}\begin{bmatrix} \cosh b & \sinh b \\ \sinh b & \cosh b \end{bmatrix}\right) = \det\left(\begin{bmatrix} t & z \\ z & t \end{bmatrix}\right) \qquad (22.5)$$

$$d^2 = t^2 - z^2$$

We have just seen the essence of the special theory of relativity emerge directly from the order two cyclic finite group C_2.

Division algebras emerge from finite groups:
There's more. We have the Euclidean complex numbers:

$$\mathbb{C} = a + ib = \begin{bmatrix} a & b \\ -b & a \end{bmatrix} \qquad (22.6)$$

$$\exp\left(\begin{bmatrix} a & b \\ -b & a \end{bmatrix}\right) = \begin{bmatrix} r & 0 \\ 0 & r \end{bmatrix}\begin{bmatrix} \cos b & \sin b \\ -\sin b & \cos b \end{bmatrix} \qquad (22.7)$$

Every division algebra (type of complex numbers) is underlain by a finite group. Above, (15.8), we saw the quaternions emerge from the group $C_2 \times C_2$.

What about the order three cyclic finite group, C_3? Simple, take the adjoint representation (Standard Form Cayley table) of C_3 and take the exponential:

$$\exp\left(\begin{bmatrix} a & b & c \\ c & a & b \\ b & c & a \end{bmatrix}\right) = \begin{bmatrix} e^a & 0 & 0 \\ 0 & e^a & 0 \\ 0 & 0 & e^a \end{bmatrix} \begin{bmatrix} v_A(b,c) & v_B(b,c) & v_C(b,c) \\ v_C(b,c) & v_A(b,c) & v_B(b,c) \\ v_B(b,c) & v_C(b,c) & v_A(b,c) \end{bmatrix}$$

(22.8)

This is the polar form of one of the four 3-dimensional types of complex numbers. The matrix with the $v_i(b,c)$ functions is a 3-dimensional rotation matrix, and the $v_i(b,c)$ functions are the trigonometric functions of this space. Note that this space is nothing like the Euclidean 3-dimensional space we see around us. For a start, there is no 2-dimensional sub-space in this 3-dimensional space.

The 3-dimensional trigonometric functions accept two arguments, $\{b,c\}$, to form the 3-dimensional angle.[47] There are four separate types of 3-dimensional complex numbers and four 3-dimensional 'complex planes' to match them. This area of mathematics is very much unexplored.

In general, taking the exponential of the adjoint representation of any finite group will produce the polar form of a n-dimensional complex number division algebra where n is the order of the finite group.[48] This polar form will have a n-dimensional rotation matrix holding n-dimensional trigonometric functions whose argument, angle, is $(n-1)$ real variables.

---

[47] For more details, see : Dennis Morris : Complex Numbers The Higher Dimensional Forms.
[48] These complex algebras are listed in : Dennis Morris : The Uniqueness of our Space-time to order fifteen.

Angles:

We have angles in our universe. Mathematicians have known about angles for thousands of years, but their knowledge was based entirely on observation. Euclid began with the fact of the existence of 2-dimensional angles and proceeded to construct geometry. The only source of angles is from within the division algebras that derive from the finite groups.

It is one thing to know that angles exist because we observe them; it is a very different thing to understand where they come from – why they exist.

We have seen above, (22.2) & (22.7) that the two 2-dimensional rotation matrices emerge from the finite group of order two, C_2. These two 2-dimensional rotation matrices hold the two types of 2-dimensional trigonometric functions which hold the two types of 2-dimensional angles. We have seen above, (22.8), the emergence of 3-dimensional angles from the order three group C_3. Angles are part of a division algebra space like the complex plane. Angles come in different types within a given dimension, and there are types of angles in every dimension. Our experience leads us to familiarity with only the two types of 2-dimensional angles we see around us.

There is no place in mathematics other than the division algebras where we find angles; they do not pop into existence just because a mathematician wants them to do so. Each type of division algebra has its own type of angle.

Summary:

The finite groups hold within them all the division algebras (types of complex numbers) and all the spaces associated with these division algebras. An aspect of each of these spaces is the angle within the rotation matrix of that space.

Chapter 23

Our 4-dimensional Space-time

Our 4-dimensional space-time is not the space of a division algebra. We do not get our 4-dimensional space-time by taking the exponential of the adjoint representation of any finite group. If our 4-dimensional space-time did emerge from a finite group, then we would have only one type of angle in our space-time and that angle would be a 4-dimensional one. We have six 2-dimensional angles in our 4-dimensional space-time. We have three 2-dimensional Euclidean angles in three orthogonal 2-dimensional planes, and we have three 2-dimensional space-time angles, hyperbolic angles, in three orthogonal 2-dimensional planes.

The distance function of our 4-dimensional space-time is:

$$d^2 = t^2 - x^2 - y^2 - z^2 \qquad (23.1)$$

This function does not maintain its form under multiplication – we do not have the product of norms equals the norm of products which we have within a division algebra. There is one too many minus signs:

$$(t^2 - x^2 - y^2 - z^2)(a^2 - b^2 - c^2 - d^2) \neq T^2 - X^2 - Y^2 - Z^2$$
$$(23.2)$$
$$(t^2 + x^2 - y^2 - z^2)(a^2 + b^2 - c^2 - d^2) = T^2 + X^2 - Y^2 - Z^2$$

We have presented an example of the product of norms equals the norm of products for comparison.

Spinning discs:
Within a division algebra space like the complex plane, there is a single rotation matrix. Within our 4-dimensional space-time, there are six 2-dimensional rotation matrices. Within a division algebra space, nothing

happens. The division algebra space just sits there outside of time. In our 4-dimensional space-time, things happen. In our 4-dimensional space-time discs can spin.

Consider a spinning disc; this disc is obviously rotating in a single 2-dimensional Euclidean plane. With thought, we realise that, as the disc spins, a point upon the edge of the disc is accelerating and decelerating in two orthogonal directions. Acceleration and deceleration is just rotation backwards and forwards in 2-dimensional space-time. We see that the disc is moving in three 2-dimensional planes; one of these is the 2-dimensional Euclidean plane and two of these are 2-dimensional space-time planes.

In a division algebra space, which has only one rotation matrix, discs could not spin. To spin, a disc needs to be sitting within a space that holds more than one rotation matrix, in fact, at least three.

2-dimensional rotations:
A rotation is a movement which keeps invariant the distance from the centre of rotation to the point moved. The 2-dimensional Euclidean rotation holds invariant the distance measured by the Euclidean distance function:

$$d^2 = x^2 + y^2 \qquad (23.3)$$

The 2-dimensional space-time (hyperbolic) rotation holds invariant the distance measured by the space-time distance function:

$$d^2 = t^2 - z^2 \qquad (23.4)$$

If we are to have the three 2-dimensional rotations needed for a disc to spin in one space, then that space must have a distance function which includes at least:

$$d^2 = t^2 - y^2 - z^2 \qquad (23.5)$$

There could be extra terms with different variables, and these extra terms could take any form, cubic, power ten if you like, but, provided that we have a distance function of which (23.5) is part, then a disc can spin in this space.

When a disc spins, three 2-dimensional spaces interact. If we had a distance function like:

$$d^3 = a^3 + b^3 + c^3 - 3abc \qquad (23.6)$$

Then there could be no 2-dimensional rotations in this space because 2-dimension rotations respect distance as measured by (23.3) & (23.4). In a 3-dimensional space with distance function (23.6), there could be no interaction between 2-dimensional spaces.

From where do spaces come?:

Spaces come from two places. One of these places is the imagination of mathematicians who, like the writers of fairy stories, simply invent a type of space. The other of these places is the finite groups.

The only mathematical source of empty spaces are the finite groups. Taking the exponential of the adjoint representation of any finite group will give us a space.

Our task is now clear. Although we take the exponentials of the adjoint representation of every finite group to form a space, the distance function of that space is just the determinant of the adjoint representation seen as a matrix. We calculate the determinant of the adjoint representation of every finite group, and we inspect that determinant to see if it contains the determinants of any finite group of lesser order.[49]

For example, the distance function, determinant, of a 3-dimensional space from the finite group C_3 is given above, (23.6). If we were to find that a 4-dimensional space, from perhaps the finite group C_4, had the determinant, distance function:

---

[49] This is done in : Dennis Morris : The Uniqueness of our Space-time.

$$d^3 = a^3 + b^3 + c^3 + x^3 - 3abc - 3abx - 3acx - 3bcx \qquad (23.7)$$

then we see that this proposed space, (23.7), can hold four separate 3-dimensional rotations. The four 3-dimensional spaces could interact in such a space; perhaps something, 3-dimensionally, could spin. There is no such 4-dimensional space with the distance function (23.7).

The uniqueness of our space-time:
Having searched through all possible determinants of the adjoint representations of all finite groups, we find, quite astoundingly, that lesser dimensional rotations can be contained in higher dimensional spaces in only one case. This case is the finite group $C_2 \times C_2$.

The group $C_2 \times C_2$ is an abelian group. It is therefore surprising that this group has within it, as well as two types of commutative division algebras, two types of non-commutative division algebras.[50] In total, there are eight non-commutative division algebras within the $C_2 \times C_2$ group.

Of the eight non-commutative 4-dimensional algebras, two are quaternion type algebras, the left-chiral quaternions and the right-chiral quaternions:

$$\mathbb{H}_{L\chi} = \begin{bmatrix} a & b & c & d \\ -b & a & -d & c \\ -c & d & a & -b \\ -d & -c & b & a \end{bmatrix} \qquad \mathbb{H}_{R\chi} = \begin{bmatrix} a & b & c & d \\ -b & a & d & -c \\ -c & -d & a & b \\ -d & c & -b & a \end{bmatrix}$$

$$(23.8)$$

The commutation relations of the right-chiral quaternions are the reverse of the commutation relations of the left-chiral quaternions. These algebras are the division algebra forms of the single Clifford algebra $Cl_{0,2}$.

---

[50] See : Dennis Morris : The Physics of Empty Space.

The other six non-commutative 4-dimensional division algebras are the A_3 algebras which are:

$$SSA^*_{L\chi} = \exp\left(\begin{bmatrix} a & b & c & d \\ b & a & -d & -c \\ c & d & a & b \\ -d & -c & b & a \end{bmatrix}\right) \quad SSA^*_{R\chi} = \exp\left(\begin{bmatrix} a & b & c & d \\ b & a & d & c \\ c & -d & a & -b \\ -d & c & -b & a \end{bmatrix}\right)$$

(23.9)

$$SAS_{L\chi} = \exp\left(\begin{bmatrix} a & b & c & d \\ b & a & d & c \\ -c & d & a & -b \\ d & -c & -b & a \end{bmatrix}\right) \quad SAS_{R\chi} = \exp\left(\begin{bmatrix} a & b & c & d \\ b & a & -d & -c \\ -c & -d & a & b \\ d & c & b & a \end{bmatrix}\right)$$

(23.10)

$$ASS_{L\chi} = \exp\left(\begin{bmatrix} a & b & c & d \\ -b & a & -d & c \\ c & -d & a & -b \\ d & c & b & a \end{bmatrix}\right) \quad ASS_{R\chi} = \exp\left(\begin{bmatrix} a & b & c & d \\ -b & a & d & -c \\ c & d & a & b \\ d & -c & -b & a \end{bmatrix}\right)$$

(23.11)

These are algebraically isomorphic. They are the division algebra forms of the two Clifford algebras $Cl_{2,0} \cong Cl_{1,1}$. All eight of these non-commutative algebras have the general form:

$$\begin{bmatrix} a & b & c & d \\ \beta b & a & \dfrac{\beta}{\varepsilon}d & \varepsilon c \\ \eta c & -\dfrac{\eta}{\varepsilon}d & a & -\varepsilon b \\ -\dfrac{\beta\eta}{\varepsilon^2}d & \dfrac{\eta}{\varepsilon}c & -\dfrac{\beta}{\varepsilon}b & a \end{bmatrix}$$

(23.12)

The determinants, distance functions, of the quaternion algebras are:

$$d^2 = w^2 + x^2 + y^2 + z^2$$
$$d^2 = w^2 + x^2 + y^2 + z^2 \tag{23.13}$$

Each of these, (23.13), will hold six 2-dimensional Euclidean rotations. The determinants, distance functions, of the A_3 algebras are:

$$d^2 = t^2 - x^2 - y^2 + z^2$$
$$d^2 = t^2 - x^2 - y^2 + z^2$$
$$d^2 = t^2 - x^2 + y^2 - z^2$$
$$d^2 = t^2 - x^2 + y^2 - z^2 \tag{23.14}$$
$$d^2 = t^2 + x^2 - y^2 - z^2$$
$$d^2 = t^2 + x^2 - y^2 - z^2$$

Each of these, (23.14), will hold six 2-dimensional rotations of which two will be Euclidean rotations and four will be space-time rotations.

Our space-time:

It would seem that our 4-dimensional space-time is six A_3 spaces taken together with two quaternion spaces sitting atop the six A_3 spaces.[51]

Of the six A_3 spaces, three are left-handed and three are right-handed.

Adding the six distance functions of the A_3 spaces gives:

$$6d^2 = 6t^2 - 2x^2 - 2y^2 - 2z^2$$
$$3d^2 = 3t^2 - x^2 - y^2 - z^2 \tag{23.15}$$

The 3 coefficient of the d^2 is a simple scaling factor which can be ignored, and the 3 coefficient of the t^2 will disappear if we choose the correct units in which to measure time. Thus we are left with the

---

[51] The mathematical details of how our 4-dimensional space-time emerges from the $C_2 \times C_2$ finite group are presented in : Dennis Morris : Upon General Relativity.

distance functions of the only spaces which can support lesser dimensional rotations. These spaces have distance functions:

$$d^2 = t^2 - x^2 - y^2 - z^2 \qquad (23.16)$$

$$d^2 = w^2 + x^2 + y^2 + z^2 \qquad (23.17)$$

We think this is why our universe is the 4-dimensional form that we observe. Although we do not discuss it at length, we have shown elsewhere[52] that both general relativity and classical electromagnetism emerge from the A_3 algebras[53]. But what about the two quaternion algebras?

Electrons and neutrinos:
Within quantum field theory, an electron is described by a quaternion written in the obfuscating notation[54]:

$$\psi = \begin{bmatrix} a+ib \\ c+id \end{bmatrix} = Quaternion \qquad (23.18)$$

We think the phenomenon of the weak force, electrons and neutrinos is how we observe the interaction between the quaternion spaces and our 4-dimensional space-time.

Summary:
There are only two types of space in which lesser dimensional rotations can exist. One of those types of space is our 4-dimensional space-time. The other of those spaces is the space of the electron and the neutrino.

So now you see why physicists ought to know about finite groups.

---

[52] See : Dennis Morris : The Physics of Empty Space.
[53] See : Dennis Morris : Upon General Relativity.
[54] See : Dennis Morris : The Quaternion Dirac Equation.

Chapter 24

Concluding Remarks

We began with nothing more than the permutations of several objects. We finished with the nature of our 4-dimensional space-time and, although we did not show any details, classical electromagnetism and electrons and neutrinos. It is remarkable that our universe falls out of nothing more than the permutations of four objects.

Along the way, we saw that sequentially combining permutations is matrix multiplication. We also discovered an enormous plethora of different sets of permutations which are the different finite groups. We found a great deal of subgroup structure within these sets of permutations. We found that there are odd permutations and even permutations and that these correspond to odd permutation matrices and even permutation matrices.

As we neared the end of this book, we found that these closed sets of permutations, which are the finite groups, underpin the various types of complex numbers, division algebras to give them their technical name. Numbers are the foundation of mathematics, and matrix multiplication is the multiplication operation by which any kind of numbers, complex or otherwise, are multiplied together. It is very sobering to realise that multiplication is really just sequentially combining permutations together.

In this short book, we have hardly put our toe into the ocean of group theory. There are great swathes of group theory upon which we have not even touched. Group theory is a huge area of mathematics wholly known to no one. Specialist group theorists spend their entire lives researching groups and see only the smallest part of this area of mathematics. Indeed, there are mathematicians who have spent their entire lives to discover only a single one of the exceptional, sporadic, groups.

Concluding Remarks

We have not gone into the technical details of proofs but have often simply stated the fact and left it unproven. Such proofs are widely available to anyone who wishes to see them either on the internet or in group theory text books. We did not want to bury the concepts in too much detail.

If the reader is new to finite group theory, we hope the reader has gained a reasonably deep understanding of the concept of a finite group from this book and that this understanding will make any future studies of group theory much easier and more enlightening.

If the reader was familiar with finite group theory before reading this book, then we hope that we have shone a light upon finite groups from an angle that has illuminated to that reader much that before was obscure or unseen.

We hope we have succeeded in 'juicing up' finite group theory. Other presentations do seem very dry not because of any failing of the authors but because finite group theory is so full of details. Most of all, we hope the reader has, at an intellectual level, enjoyed reading this book and that it will form a future reference for the reader.

Having said all of the above, you really do need to read this book three times to grasp all that is within it.

Thank you for your polite attention; it has been a pleasure writing for you.

Dennis Morris

Brotton

September 2016

Bibliography

Groups and Symmetry	M. A. Armstrong
Symmetry and the Monster	Mark Ronan
Atlas of Finite Groups	J. H. Conway, R. T. Curtis
	S. P. Norton, R. A. Parker
	R. A. Wilson

Other Books by the Same Author

Other Books by Dennis Morris

The Naked Spinor – A Rewrite of Clifford Algebra

Spinors exist in Clifford algebras. In this book, we explore the nature of spinors. This book is an excellent introduction to Clifford algebra.

Complex Numbers The Higher Dimensional Forms – Spinor Algebra

In this book, we explore the higher dimensional forms of complex numbers. These higher dimensional forms are connected very closely to spinors.

Upon General Relativity

In this book, we see how 4-dimensional space-time, gravity, and electromagnetism emerge from the spinor algebras. This is an excellent and easy-paced introduction to general relativity.

From Where Comes the Universe

This is a guide for the lay-person to the physics of empty space.

Empty Space is Amazing Stuff – The Special Theory of Relativity

This book deduces the theory of special relativity from the finite groups. It gives a unique insight into the nature of the 2-dimensional space-time of special relativity.

The Nuts and Bolts of Quantum Mechanics

This is a gentle introduction to quantum mechanics for undergraduates.

Quaternions

This book pulls together the often separate properties of the quaternions. Non-commutative differentiation is covered as is non-commutative rotation and non-commutative inner products along with the quaternion trigonometric functions.

The Uniqueness of our Space-time

This book reports the finding that the only two geometric spaces within the finite groups are the two spaces that together form our universe. This is a startling finding. The nature of geometric space is explained alongside the nature of division algebra space, spinor space. This book is a catalogue of the higher dimensional complex numbers up to dimension fifteen.

Lie Groups and Lie Algebras

This book presents Lie theory from a diametrically different perspective to the usual presentation. This makes the subject much more intuitively obvious and easier to learn. Included is perhaps the clearest and simplest presentation of the true nature of the Lie group $SU(2)$ ever presented.

The Physics of Empty Space

This book presents a comprehensive understanding of empty space. The presence of 2-dimensional rotations in our 4-dimensional space-time is explained. Also included is a very gentle introduction to non-commutative differentiation. Classical electromagetism is deduced from the quaternions.

The Electron

This book presents the quantum field theory view of the electron and the neutrino. This view is radically different from the classical view of the electron presented in most schools and colleges. This book gives a very clear exposition of the Dirac equation including the quaternion rewrite of the Dirac equation. This is an excellent introduction to particle physics for students prior to university, during university and after university courses in physics.

The Quaternion Dirac Equation

This small book (only 40 pages) presents the quaternion form of the Dirac equation. The neutrino mass problem is solved and we gain an explanation of why neutrinos are left-chiral. Much of the material in this book is drawn from 'The Electron'; this material is presented concisely and inexpensively for students already familiar with QFT.

An Essay on the Nature of Space-time

This small and inexpensive volume presents a view of the nature of empty space without the detailed mathematics. The expanding universe and dark energy is discussed.

Elementary Calculus from an Advanced Standpoint

This book rewrite the calculus of the complex numbers in a way that covers all division algebras and makes all continuous complex functions differentiable and integrable. Non-commutative differentiation is covered. Gauge covariant differentiation is covered as is the covariant derivative of general relativity.

Even Mathematicians and Physicists make Mistakes

This book points out what seems to be several important errors of modern physics and modern mathematics. Errors like the misunderstanding of rotation, the failure to teach the higher dimensional complex numbers in most universities, and the mathematical inconsistency of the Dirac equation and some casual errors are discussed. These errors are set in their historical circumstances and there is discussion about why they happened and the consequences of their happening. There is also an interesting chapter on the nature of mathematical proof within our society, and several famous proofs are discussed (without the details).

Finite Groups – A Simple Introduction

This book introduces the reader to finite group theory. Many introductory books on finite groups bury the reader in geometrical examples or in other types of groups and lose the central nature of a finite group. This book sticks firmly with the permutation nature of finite groups and elucidates that nature by the extensive use of permutation matrices. Permutation matrices simplify the subject considerably. This book is probably unique in its use of permutation matrices and therefore unique in its simplicity.

The Left-handed Spinor

This book covers the left-handed parts of mathematics which we call the chiral algebras. These algebras have CP invariance, violation of parity, and many other aspects which makes them relevant to theoretical physics. It is quite a revelation to discover that mathematics is left-handed.

Non-commutative Differentiation and the Commutator

(The Search for the Fermion Content of the Universe)

This book develops the theory of non-commutative differentiation from the fundamentals of algebra. We see what an algebraic operation (addition, multiplication) really is, and we discover that the commutator is a third fundamental algebraic operation within some division algebras. This leads to the first part of the derivation of the fermion content of the universe.

Index

2

2-dimensional Euclidean space, 77
2-dimensional space-time, 153

3

32-dimensional string theory, 151
3-dimensional angle, 155
3-dimensional angles, 156
3-dimensional complex numbers, 57, 155
3-dimensional rotation matrix, 75
3-dimensional trigonometric functions, 57, 75, 155

4

4-dimensional space-time, 157

A

A3 algebras, 161
A3, adjoint representation, 97
A3, fundamental representation, 97
A4 as a tetrahedron, 137
A4 conjugacy classes, list of, 136
A4 fundamental representation, 136
A4, adjoint representation, 97
A4, fundamental representation, 97
A4, subgroups, 106
A5, subgroups, 106
Abel, Niels Henrik, 39
abelian Cayley table, 53
abelian group, 120
abelian group, conjugacy, 130
abelian group, conjugacy classes, 130
abelian group, cosets, 126
abelian group, normal subgroups, 142
abelian group, subgroups, 105
abelian groups, 39
abelian groups of order 360, 121
abelian groups, finding them, 120
abelian groups, subgroups of, 73
adjoint representation, 49, 97
adjoint representation, A3, 97
adjoint representation, A4, 97
adjoint representation, C2, 153
adjoint representation, C2 X C2, 91
adjoint representation, C3, 70
adjoint representation, C4, 114
adjoint representation, cyclic group, 65
adjoint representation, cyclic groups, 97
adjoint representation, dihedral group, 81
adjoint representation, S3, 99
adjoint representation, S4, 97
adjoint representation, symmetric groups, 101
adjoint representations, C4, 50
algebra of permutations, 11
algebraic matrix form, 54, 57
algebraic matrix form, symmetric groups, 101
alternating group, 1
alternating group, order, 42, 93
alternating group, subgroups, 94
alternating groups, 93
alternating normal subgroup of symmetric group, 142
angles, 156
Aschbacher, Michael, 150
associative, 60
associativity, 62
Atlas of Finite Groups, 151
axioms, group, 60

B

Bocherds, Richard, 151
braid groups, 63

171

C

C2 subgroup within C4, 51
C2 X C2, 160
C2 X C2 group, 114
C2 X C2 X C2, 115
C2 X C2, adjoint representation, 91
C2 X C2, Standard Form Cayley table, 116
C3, 12, 19, 30, 33
C3, 2-dimensional representation, 33
C3, 6-dimensional representation, 35
C4, 51, 67, 70
C4 adjoint representations, 50
C6, 84
Cauchy, Augustin-Louis, 72
Cauchy's theorem, 106
Cayley table, 52
Cayley table in Standard Form, 54
Cayley table, dihedral group, 84
Cayley table, S3, 47
Cayley, Arthur, 8, 52
Cayley's theorem, 8
centre as a normal subgroup, 142
centre of a group, 107, 131
Chevalley groups, 2
circulant matrices, 66
classical electromagnetism, 163
Classification Theorem, 150
Clifford algebra, 160
Clifford algebras, 58, 116
Clifford, William Kingdon, 59
closed set of permutations, 8, 12
closure, 19, 60
Cockle, James, 154
column swap, 102
combining permutations, 17
commutator, 144
commutator subgroup, 144
commutator subgroup, normal subgroup, 144
complete set of permutations, 12
complex numbers, 116
complex numbers, matrix notation, 32
complex plane, 69, 77
conjugacy, 129
conjugacy class, 129, 132
conjugacy class, abelian group, 130
conjugacy classes of A4, list, 136
conjugacy classes of D6, list, 133
conjugacy classes of S4, list, 138
conjugacy classes, calculation, 132
conjugacy, abelian group, 130
conjugate elements, D6, 131
conjugation, cycle structure, 136
coset, 124, 142, 146
cosets of D3, list, 127
cosets, abelian group, 126
cosets, D6, 125
cosets, normal group, 142
cosh function, 40
crossed group is a cyclic group, condition, 117
crossed groups, 114, 116
cubic group, 1
cycle structure, 26, 133
cycle structure, conjugation, 136
cycle structure, symmetric groups, 136
cyclic group, 1, 12
cyclic group C4, 88
cyclic group, adjoint representation, 65
cyclic group, permutations, 68
cyclic group, subgroups, 105, 106
cyclic groups, 65
cyclic groups of prime order, 2, 67, 141
cyclic groups, adjoint representation, 97
cyclic groups, denotation, 65
cyclic groups, fundamental representation, 97
cyclic groups, properties, 68
cyclic groups, subgroup structure, 89
cyclic groups, subgroups of, 73
cyclic normal subgroups, 143
cyclic subgroup of order p, 106

D

D3, list of cosets, 127
D6, 129
D6 Standard Form Cayley table, 86
D6, conjugacy, 130
D6, conjugate elements, 131
D6, generators, 131

Index

D6, subgroup list, 110
dancing matrices, 66
determinant, 102
determinant of the adjoint representation, 159
dicyclic group, 1
dicyclic group, denotation, 112
dicyclic groups, 112
dihedral group, 1, 75, 77
dihedral group D6, cosets, 125
dihedral group, adjoint representation, 81
dihedral group, Cayley table, 84
dihedral group, conjugacy, 130
dihedral group, denotation, 79
dihedral group, form of permutations, 87
dihedral group, generators, 89
dihedral group, subgroups, 81
dihedral groups, as roots of unity, 85
direct product, 114, 116
direct product of cyclic groups, 120
direct product of simple groups, 149
disjoint permutations, 114
distance function, 154
distance function, 4-dimensional space-time, 157
distance functions, A3 algebras, 162
distance functions, quaternion algebras, 161
division algebras, 153, 156
double cover rotation, 76

E

electromagnetism, 163
electron, 163
equilateral triangle, 32, 78
equivalence relation, 133
Euclidean complex numbers, matrix form, 154
Euclidean distance function, 158
Euclidean rotation, 75
even permutations, 1, 93, 98, 99
exceptional groups, 150
exponential of the adjoint representation, 159

exponential, of adjoint representation, 153
exponential, of adjoint representation of C3, 155

F

factor group, 147
Fields medal, 151
finite group, definition, 12
finite groups, list of, 1
finitely generated abelian group, 73
finitely generated groups, 120
fundamental representation, 71
fundamental representation, A3, 97
fundamental representation, A4, 97
fundamental representation, cyclic groups, 97
fundamental theorem of finitely generated abelian groups, 121

G

Galois, Evarist, 149
Gauss, Carl Friedrich, 71
generating element, 80, 108
generating elements, 68
generating operations, 80
generator matrix, 67
generator of the group, 67, 108
generators, 88
generators, dihedral group, 89
generators, subgroups, 108
Gorenstein, Daniel E, 150
group axioms, 60
group axioms, subgroup, 104
group generators, 88
group is of prime order, 109
group of cosets, 147
groups of order 2p, 85
groups of prime order, 71

H

hexagon, 79
hexagonal plate, 129

Hölder, Otto Ludwig, 123
homotopy groups, 63
hyperbolic angles, 157
hyperbolic complex numbers, 154

I

identity element, 20, 60
identity element, denotation, 20
identity matrix, 102
identity permutation, 10, 20
identity permutation matrix, 37
index of subgroup, 126
infinite groups, 14
infinite rotation groups, 14
internal direct product, 118
intersection of two subgroups, 107
inverse, 60, 61
inverse of a permutation, 24
inverse permutation, 24
inverse permutation matrix, 35
irreducible representations, 140
isomorphism, 80

J

Jordan, Camille, 72, 123
Jordan-Hölder theorem, 123

K

Klein group, 114
Klein, Felix, 114
Klein's group, 114
knot theory, 63

L

Lagrange, Joseph-Louis, 71
Lagrange's theorem, 71, 105
Lagrange's theorem, converse, 73
leading diagonal, permutation matrix, 37
left coset, 125
left-chiral quaternions, 116, 160
Lyons, Richard Neil, 150

M

matrix multiplication, 31
monster, 151
Monster Moonshine theorem, 151
multiplication operation, permutations, 22

N

n-dimensional trigonometric functions, 155
neutrino, 163
non-abelian groups, 39
non-associativity, 62
non-commutativity, permutations, 21, 23
norm, 157
normal subgroup, 141, 143
normal subgroup, cosets, 142
normal subgroup, denotation, 145
normal subgroups, 141
normal subgroups of S3, 142
normality, non-transitive, 143

O

odd permutations, 93, 98, 99
order, 13
order eight dicyclic group, normal subgroups, 142
order of a crossed group, 117
order of a group, 13
order of a permutation, 10, 21
order of a permutation, from the cycle structure, 26
order of an element, 30, 56
order twelve dihedral group, D6, 110
order, alternating group, 42
order, symmetric group, 43

P

p- Sylow subgroups, symmetric groups, 107
partition, of group, 127
pentagon, 79

Index

perfect group, 144
permutation matrices, 6, 27
permutation matrix, 43
permutation matrix, cycle structure, 134
permutation multiplication, 22
permutation non-commutativity, 23
permutation notation, ambiguity, 10
permutation order, 21
permutation product, 21
permutation view of finite groups, 8
permutation, cycle structure, 26
permutation, defined, 9
permutation, identity, 10
permutation, inverse, 24
permutation, multiplication, 18
permutation, order, 10
permutation, order of, 88
permutation, standard notation, 9
permutations as mappings, 11
permutations, dihedral group, 87
permutations, non-commutativity, 21
permutations, quaternion group, 113
polygons, 14
polyhedron, regular, 15
polytopes, 16
powers of permutations, 20
prime divisor, 106
prime order cyclic groups, 67
prime order groups, 71
product of norms, 157
product of two permutations, 21
proper subgroup, 42
proper subgroups, 104

Q

quaternion commutation relations, 160
quaternion group, 1, 112
quaternion group, permutations, 113
quaternion group, Standard Form Cayley table, 113
quaternion rotation, 76
quaternion space, 76
quaternion type algebras, 160
quaternions, 113, 115, 160
quotient groups, 146

quotient groups of cyclic groups, 148

R

reflective symmetries, 75
regular polygon, 78
regular polyhedron, 15
regular tetrahedron, 137
representation of C3, 2-dimensional, 33
representations, 48
representations of finite groups, 35
right coset, 126
right-chiral-quaternions, 160
roots of plus one, 56
rotation matrix, 62, 155
rotation matrix, 2-dimensional space-time, 75

S

S3, 12, 29, 42, 46
S3, 2-dimensional representation, 76
S3, 6-dimensional reprersentation, 46
S3, adjoint representation, 99
S3, Cayley table, 47
S3, normal subgroups of, 142
S4, 12, 33, 51
S4 conjugacy classes, list of, 138
S4, adjoint representation, 97
Sautoy, Marcus du, 152
simple finite group, 149
simple finite groups, 2, 71, 110, 141
simple groups of Lie type, 150
sinh function, 40
Solomon, Ronald, 150
space-time distance function, 158
special relativity, 40, 153
spinning disc, 158
sporadic groups, 2, 150
square, 79
Standard Form Cayley table, 52
Standard Form Cayley table, C2 X C2, 86, 116
Standard Form Cayley table, D6, 86
Standard Form Cayley table, quaternion group, 113

standard mantra, 15
standard notation, permutation, 9
subgroup, 42, 104
subgroup generators, 108
subgroup of order p, 106
subgroup of prime order, 106
subgroup structure of cyclic groups, 89
subgroup, denotation, 104
subgroups, 42
subgroups of D6, list, 110
subgroups, A4, 106
subgroups, A5, 106
subgroups, abelian groups, 105
subgroups, cyclic groups, 105, 106
subgroups, dihedral group, 81
subgroups, list of, 1
Sylow subgroups, 106
Sylow -subgroups, 106
Sylow theorems, 106
Sylow, Peter Ludwig Mejdell, 106
symmetric group, 1, 12, 102
symmetric groups, 42
symmetric groups, adjoint representation, 101
symmetric groups, algebraic matrix form, 101
symmetric groups, cycle structure, 136
symmetric groups, denotation, 65
symmetric groups, p- Sylow subgroups, 107
symmetrical Cayley table, 53
symmetry, 75

T

T group, 112
tetrahedron, 137
tetrahedron, A4, 137
theorems, subgroups, 106
theorems, Sylow, 106
transpose, of permutation matrix, 36
triangles, 14
trigonometry, 40
Twisted Chevalley groups, 2

U

union of conjugacy classes, 141

V

vier gruppen, 114

Z

zero divisors, 58